The Reckoning
A Definitive History of the COVID-19 Pandemic and Other Absurdities

Thomas Beckett Kane

Table of Contents

Copyright   1
NOTE ON SOURCES   2
MADNESS   3
THE ROAD TO WUHAN   36
THE GOOD DOCTOR   65
SUMMER   88
MEDIA- HYPOCHONDRIATIC COMPLEX   108
CROSSING THE RUBICON   122
OUR FRIENDS ACROSS THE ATLANTIC   135
CONCLUSION   144
NOTES   159

Copyright

NOTE ON SOURCES

All assertions made in this book are supported by reputable sources. For aesthetic purposes, it does not contain a plethora of footnote numbers cluttering the text. Therefore, when possible, citations have been condensed into one footnote at the end of each paragraph. If reading the eBook version, embedded links to sources are provided where applicable. In the back, you will find a full list of notes.

MADNESS

"Evil comes from a failure to think. It defies thought, for as soon as
thought tries to engage itself with evil and examine the premises and
principles from which it originates, it is frustrated because it finds nothing
there. That is the banality of evil."
— *Hannah Arendt, Eichmann in Jerusalem*

There are episodes centuries in the past so important that debates
still rage about their meaning to this day. There are others so impactful, so
utterly shocking and profound in what they reveal about civilization, the
human condition and ourselves—that they are rarely spoken of again. Its
experience consigned to a kind of nether zone, vaguely remembered but
never addressed. A period so humiliating and revealing that its history is
better forgotten than reflected upon. A condition only natural after
someone's thin veil of civility is unmasked—their madness put on full
display. Think the Germans during World War Two or the Chinese during
Mao's Cultural Revolution. When mass movements mutate into mass
psychosis, regular individuals commit heinous acts, spurred by a sense of
justice, moral duty or an ingrained revulsion to infection. What was once
unthinkable is rendered not only acceptable but codified into law,
rationalized and made mundane.

In 1960, Israeli intelligence agents captured Adolf Eichmann while
he lived under a pseudonym in Argentina. Eichmann, a high-ranking SS
officer in Nazi Germany, oversaw the transport of European Jews to the
concentration camps. He meticulously planned the train schedules,
passenger manifests, and other logistical considerations necessary to send
millions to their death.

Standing trial in Jerusalem for his crimes the following year, court
reporter Hannah Arendt covered the scene. She described how Israeli
prosecutors attempted to portray Eichmann as a fanatical, murder-
obsessed monster driven by deep hatred toward Judaism. While Arendt—
who was Jewish—obviously condemned Eichmann's actions, she
questioned the prosecutors' caricature. In her eyes, Eichmann was not

some ghoulish villain blinded by bloodthirst but a mere functionary following orders within a bureaucracy.

In doing so, she coined the term "banality of evil" to describe how history's most heinous acts are not rooted in cartoonish evilness but happen gradually—often perpetrated by disinterested clerks filing paperwork. For example, before the Holocaust, the Nazis invited contractors to submit design proposals for their prospective death camps. Architects submitted configurations for how barbed wire would intersect with watchtowers and how pathways connected communal showers to the crematoria. What was evil had turned bureaucratic, sterile, and banal.

In many ways a more terrifying phenomenon than some kind of malevolent force surging insidiously throughout the world. Understanding vile behavior as the product of seeping, oftentimes boring processes explains how perfectly normal people can turn into frightening beasts. Thankfully, these moments are rare—the human condition a peaceful and harmonious one across the vast expanse of time. However, sometimes the veil slips, and humanity descends into madness.[1]

When U.S. health authorities fund experiments on viruses to render them more transmissible to humans, later precipitating the destruction of modern civilization—all in the name of science—you know we have descended once more. It sounds like a sci-fi movie or if real, a crime worthy of prosecution to the fullest extent of the law. What actually happened is far stranger. Instead, society celebrated these people, elevating them to the highest positions of government, even above elected officials. Their recommendations to close schools, businesses, parks, and churches hailed as gospel with skeptics smeared, ridiculed and silenced. A complete perversion of the tenets of the scientific method which prides skepticism, fallibility and humility.

Millions would die as a result, not just in the United States but around the world, especially in developing countries where disruptions of supply chains have calamitous effects. Not a single one of these "experts" nor the bureaucrats who willfully implemented their recommendations were ever charged with a crime. Instead, authorities threw citizens in jail for posting on social media, refusing to wear uncomfortable plastic cloth on their face, or committing that grievous sin of wanting to say goodbye to dying loved ones. This is the sad, bewildering but ultimately true history of the COVID-19 pandemic.[2]

From March 2020 to the end of 2022, the United States underwent the most profound and destructive period of the postmodern era. It was not a war, a holocaust, nor an environmental disaster. Instead, a group of technocrats, media professionals, and for lack of a better word—hypochondriacs—descended the world into chaos and destruction. All in a futile effort to stop the spread of a cold-like virus with a less than 0.2% fatality rate.[3]

An overstatement? Lockdowns unemployed over twenty million people in the United States alone, plunged dozens of its cities into lawlessness, and precipitated the cessation of trade and commerce globally. Previously well-equipped citizens resorted to food banks to feed their families. Millions of unemployed, purposeless, and frankly bored individuals took to the streets to destroy their neighborhoods. Suicide rates skyrocketed along with drug and alcohol abuse—not to mention pornography consumption.[4] The image of a depressed, out-of-work stoner watching porn is not a good one. In total, an estimated two million people died from lockdown implementation in the United States, far exceeding the number killed by COVID.[5]

In order to avert total economic collapse, central bankers did what they do best: print trillions of dollars—seven of them to be exact. Twenty-first-century economies were not designed to shut down for a day, let alone weeks, months or years. The economic system depends on steady flows of revenues and debt repayments day after day. Their termination meant the end of that system, forcing dramatic financial maneuvers by central banks around the world—the effects of which society confronts today through increased inflation and wealth inequality.[6]

Printed money is not uniformly distributed across the population but is injected at specific points. First introduced into the financial system, newly minted dollars are accessed by the wealthy and well-connected before anyone else. While a fraction of them do filter down to the lower and middle class, the vast majority are invested in high net-worth asset classes. Hence, the exponential rise in real estate and stock prices following the pandemic. As a result, housing prices are significantly out of reach for the average American. Meanwhile, the wealthiest castes of society enjoy an unprecedented level of prosperity, driven in large part by a money supply-fueled surge in the value of their assets.[7]

While framed as a backlash against police injustice, the riots that wracked America's cities in summer 2020 were a byproduct of the shutdowns. Unemploying millions and encasing them inside their homes is only possible for so long before law and order disintegrates. Yes, pent-up frustration towards perceived discrimination by police departments contributed, but its manifestation was only made possible by the forced unemployment and restriction of millions of Americans.

Often forgotten is why the protests subsided later that summer, coinciding with a relaxation in confinement policies. Empty platitudes hailing the virtues of defunding police departments and ritual self-flagellation of privileged castes accomplished nothing. In actuality, city governments confronting the total destruction of their tax bases by easing restrictions stemmed the unrest. Thus, lockdown-induced frustration motivated the Black Lives Matter protests just as much, if not more, than police brutality. The riots further demonstrated that our generally harmonious and peaceful society is only a hair's breadth away from total breakdown.[8]

George Floyd died in May of 2020, only two months after lockdown restrictions began. Roughly sixty days was all it took for our veneer of civility to crumble—Floyd's death serving as merely one of numerous potential triggers to ignite the mayhem. Nevertheless, thousands of small businesses never reopened and urban America accelerated its freefall into irrelevancy. Additionally, the social malaise that accompanied the breakdown of law and order only magnified feelings of nihilism and despondency within the population that have yet to dissipate.[9]

All a mighty price to pay to combat the spread of a disease with a less than 0.2% lethality rate. Of course, *The Washington Post* or Fox News will throw out a number around 2-4%. If they are halfway inclined to be honest, maybe 1%.[10] However, they are citing the Case Fatality Rate (CFR) which is the percentage of *confirmed* cases that ended in death. At the inception of the pandemic, detection for COVID was limited given the requisite infrastructure to conduct large-scale testing had yet to develop. Therefore, only those with clear signs of respiratory distress were screened. In short, the worst cases of the disease comprised the majority of testing. It does not take an Einsteinian leap of logic to grasp that the fatality rate of those particular cases is going to be very high.[11]

The true way to measure the death rate of a respiratory disease with oftentimes little to no symptoms like COVID is through the Infection Fatality Rate (IFR). The IFR determines the percentage of deaths from all cases, including asymptomatic or undiagnosed ones. A study published by the National Institutes of Health found the

IFR of COVID to be 0.23%. For those under 70 years old—only 0.05%.[12] When separated from co-morbidities, meaning other lethal factors linked to confirmed deaths of COVID, the rate falls even further. For example, those who died from cancer or in a car crash but also tested positive for SARS-CoV-2 were included in overall death counts. A lovely jiu-jitsu move performed by those who yearned to restrict their fellow citizens' basic freedoms even more. Though once the facade of pseudoscience and moral posturing is removed, the real death rate from COVID falls somewhere in line with that of influenza.[13]

Of course, an average flu year does not justify the destruction of society, so epidemiologists and journalists now cite a highly misleading metric known as "excess mortality." This number predicts how many people died *indirectly* as a result of the virus—helping make the case that COVID killed millions more than even the previously inflated number indicates.[14] Here we come to yet another jiu-jitsu move. Yes, millions died as a result of draconian *lockdowns*. Suicide rates exploded, deaths from drugs and alcohol exploded, the deaths resulting from resources needlessly diverted within hospitals to treat COVID exploded. Indeed, these were *indirect* effects of the virus. Microscopic, protein-spiked viral agents did not order the lockdowns themselves. Overzealous bureaucrats aided by fanatical journalists did.[15]

Thankfully for them, now scientists and media publications credit the deaths caused by lockdowns to the virus itself! Excess mortality figures, which depending on how crazed its attributor wants to be, can put "deaths from COVID" at anywhere between ten to fifty million! Now *The Guardian* or *Nature* or whoever wants to retroactively justify their calls to eviscerate basic freedoms can run a headline putting the total deaths from COVID in the tens of millions. For example, *The New York Times* recently placed that number at over thirty million! Orwell would smile.[16]

Once we process the blatant manipulation and total arrogance required to leverage data in this way, we realize the silver lining. They have kindly provided the number of people their policies killed! Indeed, excess mortality calculates the number of individuals killed as a result of lockdown measures. No, implementing lockdowns was not *required* based on the existence of a respiratory disease. Rather, despite the insistence of those urging restraint at the time, policymakers charged forth with their intent to turn off the U.S. economy. The deaths as a result are not "excess" in relation to COVID, but a *direct* consequence of their poor decision-making.[17]

The psychotic behavior continues. Here we come to possibly the most important point regarding the pandemic. The virus was aerosolized! Microscopic, aerosolized particles, not droplets, were the primary driver of infection. Meaning viral particles lingered in rooms for hours to days at a time. *It could not be stopped.* None of the interventions which inflicted immense amounts of pain, suffering and disruption onto society were even designed to combat an airborne pathogen. Wearing masks only protects against viruses transmitted via respiratory droplets like diphtheria and influenza. The same applies to social distancing, plexiglass barriers, and any other inane protocol recommended by epidemiologists.[18]

To repeat, the SARS-CoV-2 virus which causes COVID-19 was aerosolized, meaning viral particles lingered in rooms, buildings, and open areas for hours to days. The only way to limit the transmission of an aerosolized pathogen is to have humans never engage with each other in-person. Meaning society needed to remain in lockdown for perpetuity in order to stop transmission. Given social order collapsed in less than three months in 2020, this likely would not have turned out well. The only option (and the one inevitably taken) was allowing the virus to transmit through a majority of the population, conferring immunity along the way —known as "herd immunity."[19]

Lambasted by the press and public health experts at the time, developing herd immunity was the path society inevitably took and continues to take currently. Despite the end of lockdowns and current vaccination rates for COVID-19 of near zero, variants of the virus freely circulate throughout the global population.[20] Amazingly enough, society continues to function. Why is this? The pathogen infects millions but developed immunity crushes its effects. This was, and remains, the only known method of combatting an airborne pathogen short of the forced extinction of the human race. Given the actions of certain individuals during the pandemic, it must be emphasized that extinction was not the preferred solution.[21]

Despite knowing this (presuming they paid attention in medical school long enough to understand what it means for a pathogen to be aerosolized) experts operated with the goal of "reducing COVID transmission" via lockdowns. Yes, lockdowns reduce transmission of infectious disease. Yes, forcibly separating millions of people and confining them inside their homes reduces transmission of infectious disease.[22] This is beside the point. For example, razing all of human civilization, shutting off the consumption of energy, and stopping any and all human activity would end manmade contributions to climate change. Doing so would maximize the goals of environmentalists. However, obviously, this would incur tremendous negative effects as well. There would be dramatically fewer places to live, industrialized food and energy production would collapse—billions would die. In short, implementing these measures would result in "positive" effects, but its *overall* impact would be disastrous.[23]

The same goes for public health. In theory, authorities could separate individuals from each other, lock them in quarantined bays and reduce the transmission of all infectious disease to zero. A goal of any public health department is to "reduce transmission of disease."[24] Thus, such a policy would maximally realize that goal. Of course, economic productivity would plummet, rates of depression would skyrocket along with alcohol and drug abuse, and the world would become a deeply dark and bleak place with very little to live for. Sound familiar? In any case, this was the logic followed by health and media professionals during the COVID-19 pandemic. Any and all measures were justified under the gospel of "reducing transmission," without any consideration of their costs. The complete destruction of American society deemed acceptable if doing so meant a reduction in cases.[25]

As a result, "progress" against the virus was measured in number of new cases. For lockdown enthusiasts, if that number increased in a particular state or country, then that respective government was "failing" in its response. Naturally, they demanded harsher lockdown measures. This kind of logic may have made sense if the virus exhibited a higher fatality rate. However, in the overwhelmingly large portion of cases, patients either never had any symptoms or recovered quickly.[26] Of course, when a government—whether local or national—did attempt to open up, however gradually, to alleviate the immense economic and social strains of lockdowns (in other words, doing their job as elected leaders to serve their constituencies), choruses of "such and such leader is endangering their people" inevitably rained down.[27]

Yes, by opening up, cases of an aerosolized virus with high transmissibility are going to increase. Yes, cases would remain lower if everyone remained locked in their homes. Again, this is beside the point. Elected leaders are obligated to consider a range of factors when making decisions. For example, an environmentalist walks into the office of a city mayor and demands that all available actions be taken to save the environment. If taken literally, the mayor would ignore his other constituencies, focus exclusively on the interests of environmentalism and order the entire city razed to the ground.

While ridiculous, this is not far from what most government leaders did when public health officials demanded action be taken to curtail the spread of COVID-19. As a result, public policy was directed to addressing one particular concern and not implementing a balanced plan of action to accommodate all interested parties, otherwise known as—the job of elected officials.

"But we had to lockdown the country to avoid overwhelming our healthcare system!" Nonsense. This refrain was among the most prevalent—and most powerful—responses to those calling for an end to lockdowns. Powerful in the sense, like nearly every rebuttal to those asking that their basic freedoms be restored, that it mixed just the right about of moral grandstanding with vague appeals to practicality—the ultimate form of insidious expression. If patriotism is the last refuge of scoundrels, appeals to morality are a close second.[28]

Nevertheless, even a cursory examination of reality indicates the U.S. healthcare system possessed more than enough capacity to handle the total volume of respiratory infections. Nowhere was this clearer than the near vacancy of the USNS *Comfort* dispatched to handle the supposed overflow of COVID patients in New York City. The Navy hospital ship, equipped with a thousand beds, sat nearly empty its entire time docked off the east coast of Manhattan. Similarly, hospitals around the country reported occupancy rates averaging around just forty percent. In short, the country was never in serious danger of exhausting its medical facilities nor bed space.[29]

Did some hospitals struggle to handle the influx of patients from a novel coronavirus? Of course. Does their experience justify the isolation, confinement, and immobilization of the entire U.S. population? When a traffic jam occurs in New York City, should we ban citizens from driving all around the country? Local problems do not require national solutions.

***

Nietzsche once wrote, "Madness is rare in individuals—but in groups it is the rule."[30] For over two years a sizeable portion of the population, unfortunately many of whom occupied positions of power, proved that rule. As an example, famous intellectual and renowned brainiac Sam Harris said he preferred more children die from COVID-19 because it would boost support for lockdown measures. When someone who would ostensibly score highly on an IQ test advocates for the deaths of children, you know society has gone beyond the pale.[31]

This illustrates the limits of expertise and, unfortunately, even intelligence. Thankfully, children have extremely powerful immune systems, ensuring their death rate from COVID remained less than 0.01%. Any death of a child is a tragedy but unsecured swimming pools and firearms are far greater threats to their safety than a cold virus.[32] Nonetheless, authorities proceeded with shuttering the nation's school system, precipitating lasting speech disorders and learning disabilities.[33]

What explains this reversion to insanity? An evolutionarily encoded revulsion to infectious pathogens certainly played a role, but so did darker elements of the human condition. Sadly, its instinct toward control and coercion resurfaced on a massive scale. It resurfaced in the previously cordial gym attendant who, empowered by the strictures of his local health department, began pacing the weight room yelling at guests to "put on a mask!" It resurfaced in the city employee who after being granted unprecedented powers to reduce the transmission of COVID, began issuing blanket decrees resulting in the unemployment of millions.[34]

Steeled by the mandate of "saving lives" and "protecting the vulnerable," modern society revealed itself to do anything and hurt anybody. The tender threads that bind communities in decency and respect fell apart, exposing the true nature of those around us. We need not look to a Harvard study nor a psychology paper to know that when granted authority certain individuals unleash a torrent of tyrannical behavior. Thus, while leaders at the highest echelons of power are most to blame for the COVID calamity, unfortunately part of the responsibility lies at the feet of regular citizens. After all, many of the guards at Auschwitz, Buchenwald, and Dachau, screaming at cowering mothers and children were teachers, cooks, or lawyers in their past lives.

The utter derangement stretches even further. A definitive sign humanity truly did descend into madness was the collective refusal to think logically about the origins of the virus. As recently as March 2025, *The Washington Post* reported that "there is no evidence SARS-CoV-2, the virus behind the pandemic, was in any laboratory before the outbreak." A blasphemous lie that requires the simplest of logical deductions to refute.[35]

In late 2019, there were three laboratories in the world studying novel coronaviruses. There was only *one* studying how to render the virus more transmissible to humans—in a practice known as gain-of-function research.[36] During that same period, a novel coronavirus began infecting humans in the same city in which this *one* laboratory in the world was conducting gain-of-function research on coronaviruses. That laboratory was, of course, the Wuhan Institute of Virology. Officials within the Wuhan government first began observing the outbreak of COVID in late 2019. *Imagine that!* Incredibly, then and even now, suggesting that the virus originated from the city's laboratory is considered a "theory" or even more damningly, a "conspiracy theory."[37]

Making this obvious connection, one not above the abilities of a third grader, is not to suggest the Chinese deliberately leaked COVID. First off, there was no way the Chinese Communist Party could have known how buffoonishly and self-destructively the West would react. More conclusively in rebutting this legitimate conspiracy theory is that the CCP faced a near existential threat to its authoritarian rule as a result. After almost three years into their government's failed "Zero-COVID" policy (as again, the virus was aerosolized and unstoppable), cooped up and demoralized Chinese citizens began agitating for regime change—presenting the CCP with its greatest threat to control since the Tiananmen Square protests. If China's leaders did nefariously release a novel coronavirus to boost their position relative to the West, they did a lousy job. In all likelihood, the highly transmissible virus, engineered to transmit to humans, accidently escaped and transmitted into the Wuhan population.[38]

Not surprisingly (to those familiar with the state of political discourse at the time), stating this obvious fact was rebutted by appeals to our differing skin colors. That somehow drawing the connection between the coronavirus lab and the emergence of coronavirus near the lab was casting racial aspersions against the Chinese. Another moral Hail Mary that amazingly landed as legitimate. Of course, the virus did originate from the Wuhan laboratory and rather than motivated by racism, saying so is motivated by common sense.[39]

As Jon Stewart famously said, if there was an outbreak of chocolatey goodness in Hershey, Pennsylvania, we would look to the town's local chocolate factory as the likely culprit. In similar fashion, if someone punches you in the face, your first suspect as to who punched you in the face will be the owner of the fist that connected with your face. Readers far into the future who did not experience the pandemic first-hand may feel perplexed that the origins of COVID were ever in question. Someday social psychologists smarter than this author may explain the phenomenon—but for now, we must attribute it to that ever-present, ever-lingering force: *madness.*[40]

What is undoubtable are the reasons why some wanted to push such a narrative. Funneling taxpayer money to experiment on making a virus more transmissible to humans—an outcome societies normally do *not* want to happen—to then have that same virus escape into the public is an embarrassing, likely criminal, offense. What would be even more damning would be if the *de facto* leader of the response to such a virus was the one who enabled its research. Again, it sounds like the plot to a bad sci-fi movie. Instead, meet Dr. Anthony Fauci, who as head of the National Institute of Allergy and Infectious Diseases (NIAID) was granted near complete powers to recommend (but practically order) any and all measures to stem the transmission of COVID. Revered by the press, beloved by hypochondriacs everywhere and emboldened by the all-knowing mantle of "science," Fauci pushed for the shuttering of nearly all human activity outside American homes.[41]

What he failed to mention in any of his numerous daily press briefings from the White House, as a complacent president stood behind him, was that his agency commissioned taxpayer funds for the Wuhan Institute of Virology to conduct gain-of-function research on what would become COVID-19. Indeed, beginning in 2014, NIAID allocated millions of dollars for gain-of-function research on coronaviruses to increase their transmissibility to humans. As an aside, under his leadership the agency also spent $424,000 of taxpayer funds to observe the effects of beagles being eaten alive by flies. Fun stuff, all paid for by American tax dollars.[42]

Despite documentary evidence to the contrary and eventual admittance by the National Institutes of Health, Fauci repeatedly denied any involvement to COVID gain-of-function research, even under oath to Congress. Email logs also reveal how Fauci smeared those connecting COVID to the Wuhan lab in press and scientific publications.[43]

Normally, society does not allow the perpetrators of a problem to spearhead its solution to that problem. Should burglars monitor your home security system? Should Bernie Madoff have been placed in charge of the Securities and Exchange Commission? Should Diddy run Child Protective Services? Nonetheless, Fauci took on almost God-like status during the pandemic with his word dutifully hailed as doctrine.[44]

***

It seems every available method, short of actual appeals to rationality and logic, was deployed to refute arguments against lockdowns. Racism, twisted morality, distorted statistics, outright lying and whatever else could be invoked to justify subjugation were desperately lobbed into the public discourse.

One such tactic was to distort legitimately held views, tying them to bizarre and outlandish theories. There will always be a small, but vocal minority of either attention-seeking or honestly delusional individuals intent on voicing opinions completely at odds with reality. 9/11 conspiracy theorists, Russian collusion advocates, or even those lobbying for lockdown measures and mask-wearing to this day (they do exist), come to mind. One group during the pandemic, only slightly more disconnected from reality than those clamoring for lockdowns, claimed COVID was a preplanned, government plot to enslave the American citizenry—known as "plandemic." The idea gained traction in ultra right-wing circles and in minds suffering from schizophrenia.[45]

The complexities of reality tend to bewilder conspiracy theorists. Insecure about their lack of control over world events, they feel helpless in their wake. Instead of grappling with those complexities, and possibly taking action to intervene, conspiracy theorists attribute the course of history to a cabal of faceless elites insidiously pulling the strings of global politics. Believing this idiocy rids the individual of responsibility and the guilt of inaction as *"the world is controlled by elites so what could I possibly do?"* The whole idea of reptilian Jews or whatever delusional caricature of the day is a Freudian projection of frustration on behalf of the conspiracy theorist toward their hierarchical position. In fact, the reptilian elites are just as clueless as the common man and look no further than the pandemic as evidence.[46]

Moreover, the ever-increasing complexity of society, amplified by the internet and globalization, has rendered the modern world so interconnected that even the greatest genius could not predict nor plan an event as intricate as a global pandemic. The self-immolation of one Tunisian man in 2010, followed by a mass movement on social media, sparked a regional revolution known as the Arab Spring—toppling governments and dictators. Could any Illuminati, WEF, Bohemian Grove-attending overlord possibly have predicted this or mapped out its butterfly effect? Obviously not, and the same applies to a "plandemic."[47]

The idea would be unworthy of even addressing if not for the clever co-option of the phenomenon by pro-lockdowners. Rationally minded people asking that modern civilization not be obliterated because of a new cold virus, were repackaged as claiming "COVID is not real." None of whom believed a microscopic, protein-spiked virus that attacks respiratory cells within the human body *literally* did not exist. Rather, they expressed the observation that the annihilation of our way of life was not a commensurate response to its mere presence. Nonetheless, the conspiracy theories (it is debatable whether anyone actually believed some of them) were draped over those questioning lockdowns. Sadly, these rhetorical submission holds often worked, consigning calls to liberate society to the fringes of acceptable discussion.[48]

Here, we arrive at another critical point. Neither those who attempted to defend civilization against destruction at the time nor this book argues that COVID-19 was not a legitimate threat to be addressed. People died as a result of contracting the virus. In most cases, their immune system was unable to destroy the virus before it penetrated deeply into their respiratory tract. Victims' final days were wracked by extreme pain and suffering as their lungs gradually ceased to function.

COVID robbed these individuals of the remaining years of their life—losing out on spending time with grandchildren, exploring the world, or enjoying the everyday pleasures of this already brief and fleeting existence.

In densely populated areas with a sizable number of individuals aged seventy years or older, significant social distancing measures were not only justified but absolutely necessary. This demographic is at extreme risk from dying of respiratory infection due to their weakened immune systems. Nursing homes and assisted-living facilities are venues where lockdown measures contributed significantly to the preservation of life. Additionally, a municipality like New York City, both densely populated and featuring a high proportion of elderly residents, was aptly suited to lockdown provisions.[49]

That being said, the country stretches outside New York City and geriatric care centers. The costs of shutting down localities like Topeka, Kansas; Sacramento, California; and Canton, Georgia; did not justify reducing transmission of COVID. It is a central thesis of this book that COVID-19 was a retirement home-New York City problem to be addressed by local officials familiar with conditions on the ground. *COVID was not a national problem*, requiring blanket solutions from epidemiologists lecturing from the White House press room.[50]

Another depressing byproduct of this hyperreaction is its impact on future epidemics or pandemics. Recent polling indicates a majority of Americans disapprove of how governments responded to the COVID-19.[51] Yet, a highly transmissible virus with increased lethality emerging in the near future is entirely possible. While a virus with an extremely high fatality rate *and* high transmissibility is unlikely given the pathogen would kill its hosts before it has a chance to spread widely (e.g. Ebola), it is undoubtable that someday the United States will face a serious pandemic.[52] However, the recent hysterical response has cast a long shadow. Taking requisite measures like social distancing or mask-wearing (if the virus only transmits through droplets) are less likely to be taken seriously the next time. Public health officials likely wasted their only chance to motivate citizens in combatting a real civilization-threatening pandemic like the Black Plague or Spanish Flu.

It is also essential to understand that a pathogen with x death rate in a given population will not necessarily exhibit that same death rate in another population. COVID had a different fatality rate among eighty-year-olds on iron lungs than twenty-year-old track runners. Despite this, viruses are nearly always reported as having uniform death rates.[53]

For example, avian flu has a "60% death rate" among humans. Whenever there is an outbreak among chickens, the press salivates at the opportunity to report this "fact" about the virus. Doing so implies it will kill 60% of all humans who contract it. This is nonsense. If a widespread outbreak were to occur, humans would take necessary precautions to not only avoid transmission but treat symptoms. Furthermore, as already stated, if a virus is extremely lethal then it will kill its host before it can spread widely.[54]

Nonetheless, a sure-fire way to generate attention is to publish a story about an obscure virus in sub-Saharan Africa or Southeast Asia killing the majority of an extremely small sample size of people.[55] Both the paucity of quality healthcare in those regions and its miniscule sample size will make for an alarmingly high death rate—far higher than what populations in the West would endure. Yet, similar viruses will continue to emerge for the rest of human history, providing a never-ending reservoir of eye-popping headlines for the media to pounce on. The question will be whether populations capitulate to their inane screechings once more and permit the destruction of their societies as a result.

Possibly even more depressing is how senseless recommendations by epidemiologists during COVID eroded confidence in expertise itself. A healthy skepticism toward authority is prudent but experts are often the only ones equipped to address problems and recommend solutions. Even the most anti-establishment, skeptical cynic would prefer an experienced surgeon operate on his body rather than a stockbroker with a scalpel.[56]

That being said, the virtues of science, medicine, and specialization sustained massive damage during the COVID-19 pandemic. Hordes of epidemiologists, immunologists, virologists and other "experts," normally confined to brain numbingly boring jobs in sterile laboratories, realized they could appear on CNN, be quoted in *The Washington Post,* or put on the cover of *InStyle* magazine by advocating for nationwide quarantines. Years of medical school, thousands of dollars in tuition, and countless hours staring through a microscope suddenly became worth it, if their "expertise" could be leveraged to literally control the world. Not surprisingly, trust in those fields has never been lower as a result.[57]

However, data modeling, the field responsible for projections into the future—whether it be climate, economic, or epidemiological—took arguably the biggest blow to its reputation. Already a dubious proposition given the enormous complexities of the modern world, models are notoriously inaccurate and unable to account for unforeseen influences. Often designed by the alarmist and fatalistic, they hardly ever account for human ingenuity—forecasting the most dire and dreadful outcomes imaginable.

The original prognosticator of the sort, eighteenth-century economist Thomas Malthus, infamously predicted population growth would outstrip food production, leading to catastrophic levels of death and suffering. More recently, Paul Ehrlich predicted the same, only to have food production explode via the Green Revolution in the late 1960s. In other words, we turned out all right.[58]

Nonetheless, the doomsdayers reasserted themselves in the early days of the pandemic. Academics published dozens of models, immediately seized on by the press, stating millions would die if governments did not react authoritatively. In March of 2020, the Imperial College of London released the most infamous one of all, predicting the deaths of over two million people in the United States absent lockdowns.

Devised under the leadership of Neil Ferguson, who like so many others later violated the social-distancing measures he advocated for (this time to have an affair with a married woman)—the model heavily influenced subsequent lockdown policy in the U.S. and Great Britain.

Naturally, his projections were dreadfully wrong as they assumed zero behavioral changes by the population and grossly overestimated the disease's lethality. Nonetheless, a craven media searching for yet another fearmongering headline and a public health establishment eager to assume evermore authority, pounced on Ferguson's findings.[59]

Merely being the product of a computer system and backed by the name brand of a prestigious university does not confer Nostradamus-like predictive powers on a model. If the inputs are erroneous then its outputs will be as well. Yet, in the chaotic, fear-driven early days of the pandemic, publishing a study predicting *"actually humans will adjust and this won't be that bad"* was not going to gain traction in the press. In this way, the media and academia entered into a positive feedback doomloop to only publish and promote alarmist information about the virus. Sensible data cautioning restraint and prudence was muffled with the most sensationalist and dramatic prognostications platformed. While not a new phenomenon, as newspapers have chased titillating headlines for centuries, the pandemic functioned as a uniquely potent source of fuel for the media-industrial complex to leach off.[60]

***

If dubious data models were jet fuel for the fourth estate, then "number of new cases" was dynamite laced with rocket fuel draped in kerosene. 24/7 news networks ran a continuous, upward-ticking box at the corner of their broadcasts, denoting the number of confirmed COVID cases. Curiously, the boxes fell off their screens following the 2021 presidential inauguration, but the damage had already been done. Running headlines like "U.S. Coronavirus Cases Rise by 40,000" or "U.S. Coronavirus Cases Surge Past Summer's Records" functioned to convey the virus was out of control and more stringent lockdown measures were necessary.[61] However, if 99% of cases are benign and patients recover within days or weeks, then each new case does not justify the shuttering of society. Respiratory viruses circulate through populations every year, yet the media does not incessantly publish articles chronicling their spread. In the 2023-24 flu season alone, the CDC reported twenty-eight million cases of influenza with over 300,000 hospitalizations![62]

"But COVID was a 'novel' coronavirus! We had to shut everything down until we understood what was going on! Restricting the freedom of millions was justified!" More nonsense. The truth is that COVID was nearly identical in makeup to previous novel coronaviruses such as Severe Acute Respiratory Syndrome (SARS). The two share the same genus group, making them nearly identical in genetic and structural composition. The virus which causes SARS is SARS-CoV-1. The "1" only suffixed to its name following the emergence of SARS-CoV-2.[63]

Yet, the world did not descend into a fear-induced paralytic hellscape during the 2003 outbreak of SARS. Affected populations took reasonable precautions like washing their hands more frequently and treating symptoms with over-the-counter remedies like vitamin D and zinc. Even ultra-authoritarian governments like China, where SARS first surfaced, took relatively passive approaches to the virus. The Chinese Communist Party isolated affected groups and promoted stronger public health practices whilst permitting their society to function relatively the same— nothing like their response seventeen years later.[64]

Why is this? The answer: Real-Time Polymerase Chain Reaction (RT-PCR) testing. Rarely cited alongside the most influential technologies of the age, few innovations have done more to decimate society than RT-PCR. Further discussed in Chapter Two, in essence, RT-PCR greatly reduced the time and expenditure required to discern positive cases of respiratory disease. Not yet available in 2003, testing for SARS entailed time-consuming and expensive processes dependent on centralized laboratories staffed by trained professionals—for one, single test.[65]

Whereas the advent of RT-PCR by 2020 enabled the testing of hundreds of samples at once, all in less than thirty minutes, at a fraction of the cost. Someone slightly under the weather or wanting to take time off work or simply curious to see if they had COVID, could now drive or walk to a makeshift point-of-care testing facility, and at zero cost in most places (as governments subsidized the practice) could test themselves.[66]

A technological marvel to be celebrated, right? Not quite. Its main innovation was to terrorize the public into submission by providing millions of new data points for media publications and public health officials to scream about. Nearly all of these new cases were benign, with patients experiencing little to no symptoms—but a new case is a new case, and they were dutifully logged into the ticker boxes of *The New York Times* website or at the corner of a CNN broadcast.

While confirmed cases of SARS plateaued at 8,000 cases worldwide in 2004, millions of new COVID cases are reported every year—totaling over seven hundred million. Was SARS 88,000 times less transmissible than COVID, despite sharing the same genus? Did public health authorities execute the most effective virus containment strategy in human history for SARS, an ability they apparently lost seventeen years later? Or did RT-PCR dramatically increase the detection capacity for respiratory disease, enabling millions to confirm they had a particular strain of the common cold—creating a super pipeline of data from testing facilities to public health institutions to be put on blast by the media? May history be the judge.[67]

***

Lamentably, to fully understand the pandemic, one must address the unfortunate X-factor of the early-twenty first century: Donald Trump. The dominant force of the era, no single individual is more polarizing and impactful than the former reality TV show host.

Sadly, Trump represents one of those few historical figures whose presence precludes rational debate. Rather, he induces a curious case of mass psychosis in certain segments of the population, particularly the press. A balanced appraisal is nearly impossible as one cannot acknowledge a positive facet to his presidency without being branded a right-wing MAGA zealot.[68]

While there have certainly been negative effects to his political existence, the fact remains that the president was cruising to re-election prior to the pandemic. Buoyed by a combination of tax and regulation cuts, the American economy boomed into late 2019.[69] Granted, three years of Trump, but also media and opposition-induced mayhem, understandably left the country fatigued, giving Democratic contenders a fighting chance. However, polling indicated a substantial lead for the ex-businessman headed into 2020.[70]

A desperately hostile press, keen to glom onto anything to attempt his ouster, pounced on the respiratory virus from Wuhan. The president, whose *modus operandi* is to combine his own brand of bluster with Reaganite economics was ill-suited to combating nanoparticles freely circulating through the air. Yet, his pathology to take credit for solving every national problem deceived him into seeing an opportunity to "save the day" and net another "victory."[71]

As a result, the aerosolized, highly transmissible virus—immune to any intervention sans herd immunity—functioned as the perfect kryptonite to his presidency. Indeed, no amount of ventilators, no amount of masks, no amount of Lysol was ever going to make a difference.[72] Instead, recognizing the true nature of the threat, Trump should have delegated responsibility to local authorities to address COVID at their discretion. Though, enamored by Fauci's glamorous reputation and being a septuagenarian germaphobe in his own right, Trump abdicated his duty and let the "experts" run the show. While alarmists like Fauci caused the resulting chaos, ultimately the blame for why he was even in an authoritative position lies with the commander-in-chief.[73]

Of course, delegating responsibility to state and local governments predictably spurred the press to accuse the president of "throwing up his hands" and "doing nothing." *"How can you sit back and let people die!"* they screeched. The reaction is symptomatic of larger trends in the United States, beginning during the Great Depression, to expect federal officials to solve every conceivable problem.[74]

Originally designed so that local governments and their constituencies could address concerns as they saw fit, the Founders codified decentralized decision-making into the Constitution. A virtue which helps explain the success of their political experiment. In Federalist Paper No. 45, James Madison wrote that the powers delegated by the founding charter to the federal government are "few and defined," whereas to the states they are "numerous and indefinite."[75] Before the usual suspects screech that the Constitution is "outdated" and that "times have changed," they need only look to recent assaults on freedom of speech or the erosion of safeguards against police brutality to understand why the concerns of 1789 are still relevant.[76]

Nonetheless, today, at the federal level, and especially in the executive branch, it is close to political suicide to relinquish responsibility and trust state officials to adequately address crises. Voters and the press expect national politicians to divert funding, make public statements, set up new bureaus and bend over backwards to confront any and all difficulties.[77]

For example, the enormous complexities for why economies rise and fall are now attributed to actions by the president. "Obama fixed the economy!" "No, Trump did." These ridiculous simplifications only serve to centralize control even further. When in reality, the economy of a developed, industrialized nation is dependent on innumerable other factors drastically outside the scope of any one man or woman. Simply injecting billions in Keynesian stimulus or cutting taxes can neither "fix" nor "destroy" it.[78]

Nonetheless, politicians take credit when incomes rise and blame their predecessors when they fall. Citizens expect leaders to constantly take action and alter anything at odds with their vision of utopia. In reality, inaction is often the best solution, whether in economics or fighting infectious disease.

Markets bereft of government subsidies or price controls have consistently been proven to allocate resources in the most efficient and cheapest way possible. This is not an ideological "free market, laissez-faire" opinion—but a fact of how supply and demand curves meet on a price graph.[79] Negative side-effects like pollution or inequality, known as "externalities," do materialize as a result, which legislators and state authorities can address through reasonable regulation. However, the basis for strong economic systems remains the liberalized movement of capital and labor to meet consumer demand. Look no further than the hyperinflation of Venezuelan currency or the supply shortages of the Soviet Union as evidence.[80]

The same principles apply to the successful management of infectious disease. During a pandemic, society must be permitted to operate as normally as possible with individuals taking precautions at their discretion. The majority will develop immunity reducing the virus's ability to spread widely. To protect the minority who are at serious risk of dying (the elderly and those with pre-existing conditions), state officials should intervene to safeguard their health while allowing the rest of society to function.

Arizona Governor Doug Ducey and Georgia Governor Brian Kemp implemented this strategy and saw a lower death rate from COVID than other states who embraced lockdowns—all while keeping their economies up and running.[81] Not suprisingly, lockdown-enthused media outlets savaged the two governors for doing so, scaring away others from doing the same. Even President Trump criticized Kemp's decision. The result being that suffering from the shutdown of humanity continued longer than necessary.[82]

***

This author is under no illusion of the nature of politics. The game is ruthless and opposition parties, right or left, will blame the ruling party for practically anything that occurs on its watch. Nor are the long-standing biases of the media or myth of "objective" journalism lost on yours truly. Journalists have consistently allowed their personal views to infect their work, with media outlets serving as mouthpieces for particular political parties for centuries.

During the presidential election of 1800 between John Adams and Thomas Jefferson, the pro-Jeffersonian newspaper, *Callender*, published an editorial accusing Adams of being a hermaphrodite.[83] While not as debased but just as inaccurate, modern media outlets incessantly published stories claiming the President of the United States was a Russian spy. They even awarded one a Pulitzer Prize for the effort! Writing a history of this age-old dynamic does little to further public discourse.[84]

However, a history of the COVID-19 pandemic stands to reveal so much more. Events from 2020 to 2022 exposed the illusion of society riding a smooth, upward trajectory towards more insight, more clarity and more informed decision-making. The pandemic unraveled the idea that increased understandings of microbiology, virology, and all the myriad other advances in science have made us better equipped to respond to crises.

Instead, the segmenting of a kind of cognitive elite, armed with prestigious-sounding degrees in medical and scientific fields, created an overeager and out-of-touch technocracy intent on proving its expertise. Major advances like the discovery of cells, viruses and antibiotics were made centuries ago leaving public health one of the last domains for ambitious Masters and Ph.D. graduates to make their mark. Again, hundreds of thousands of dollars spent or borrowed for years upon years of schooling had to be put to use somehow.

Therefore, what follows is not merely the history of how an instinctual revulsion to infectious disease was exploited to catapult a cold virus into a civilization-ending level threat. Nor is it about a complicit media weaponizing those fears to terrify the country into submission; or how the virus enabled regular people to dig into that exquisite, primordial desire to rule over others and then bury it like it never happened.

Rather, this book illuminates the pitfalls of modern society and where it can stumble. How the magnificent systems erected to operate this world can cannibalize themselves. How the marvels of high-speed communication can mutate uncontrollably, enabling a small minority to descend the world into chaos and despair. At its core, it is a warning for the future.

***

In his well-known and oft-cited Farewell Address, President Dwight Eisenhower cautioned against a looming military-industrial complex dominating public policy for its own ends. While certainly valid, a lesser-known part of his speech addressed the equally disturbing prospect of a "scientific-technological elite" accomplishing the same. Operating under the guise of "expertise" and "science," this cabal of technocrats would co-opt the levers of democracy to further their own narrow fields of interest at the expense of the general public.[85]

Admittedly, advances in science have catalyzed unprecedented increases in wealth and prosperity, solving previously thought intractable problems and raising the standard of living for all humanity. As a result, scientists have deservedly taken an authoritative position in modern society. Their rise explains declines in religious belief with increasingly more phenomenon explained by their discoveries.[86]

In many respects, scientific inquiry has completely mapped out the mechanisms behind the physical world, discovering how both massive and extremely small things operate and interact. The last undiscovered frontiers, and therefore the last refuges of religion, concern the origins of all matter (how something came from nothing) and the mind-body relationship.[87] Aside from that, science has claimed total victory and understandably capitalized on its ascendance. Undoubtedly entitled to opine and advise on developments within their specific field, the opinion of experts should be valued. As said before, anyone should want an experienced surgeon to operate on them compared to an athlete or musician.

However, while expertise may provide depth it often sacrifices an understanding of the bigger picture. Recent advances in artificial intelligence evidence the point. AI developers celebrate the potential of computers to substitute human work and endeavor. To an AI researcher, this portends a glorious future and all available resources should be directed to its realization. Purposeless humans will be given an annual income and upload their minds into a simulated world to indulge in hedonistic fantasy for eternity.[88]

Yet, most people derive sense of self, identity, and even pleasure from their job. Some will even resonate with the feelings of immense boredom, emptiness, and anxiety that accompany "relaxing" or "doing nothing" for extended periods of time. Unemploying and stripping purpose from billions of people's lives would be disastrous (See: urban America, summer 2020). Instead, hype surrounding artificial intelligence more closely resembles an earlier generation of technocrats prophesying utopian visions of an interconnected populace via the internet—only to have it magnify the darker shades of the human psyche through social media and pornography.

This is not to adopt a Luddite, Kaczynskian, anti-technological creed. Advances will march on regardless and only quixotic fools oppose them completely. However, trusting specialists to steward development to the betterment of society is equally misguided. After decades of working in a particular field, experts understandably enter into a tunnel vision, overemphasizing the stakes and considerations of their own field at the expense of others. In the case of the pandemic, while not every lockdown skeptic possessed a Ph.D. in Virology or Epidemiology, they were better equipped to balance the considerations of their localities than unelected technocrats.

Nonetheless, sixty years after Eisenhower's pronouncement, the "scientific-technological elite" seized the mantle of balancing those considerations. Biology and pre-med majors were handed responsibility for addressing a problem that intersected with economics, culture, business, and practically every other dimension of human activity. It was equivalent to recognizing that artillery shells and baseballs correspond to the laws of physics, thus entrusting Neil deGrasse Tyson as Supreme Allied Commander of Europe or the manager of the New York Yankees. This is practically what society did in emplacing narrow-minded experts in charge of public policy for two years.

In homage to Eisenhower's prescient remarks and for lack of a better all-encompassing term to describe the disparate groups who advocated for lockdowns—this entity will be referred to as the "media-hypochondriatic complex."

***

Before proceeding, a few things must be noted. While arguably polemical, this is not a manifesto. Hacks write manifestos. This is a history. Unfortunately, the field of history has been bastardized as a form of chronology, confusing it with the memorization of dates and battlefields. Yet another casualty of society's recent obsession with prioritizing STEM-related fields.

Rather, its true meaning derives from the Greek word (historía), meaning "inquiry."[89] Like scientists or mathematicians, historians search for truth—the difference lies in where they look. Scientists observe the physical world, mathematicians seek out formulas that govern reality, and historians examine the past. They do so not to determine mundane factoids like which year Napoleon won a certain battle or which Sultan put down a given rebellion. Instead, true histories garner insight into life's ultimate questions. What is human nature? What determines success or failure? Does faith, kindness, and love hold any real influence or are we all just self-interested brutes fighting for ourselves? The answers, or more precisely, the interpretations historians come to as a result, shape beliefs about the present and future.

Yet, most inquiries fail to rise to this standard. Someone can write dozens of books arguing why the Japanese attacked Port Arthur during the Russo-Japanese War, a possibly interesting subject (depending on how much of a military history nerd you are), but its implications do not matter much today.

Instead, there exists a small collection of past events that mold widespread convictions in the public consciousness. Depending on whether you believe Christian missionaries aggressively threatened Islamists during The Crusades or vice-versa, likely influences how you view modern relations between the two religions. If you believe American colonists justly conquered lands held by natives, then you probably view the United States positively. How you weigh slavery versus states' rights as the cause of the Civil War helps form your opinion about the legitimacy of Southern secession. Thus, the power of history lies in its ability to leverage evidence to support one particular interpretation over another.

Of course, the most important historical episode to contemporary society is World War Two. Look only to the prevalence of Nazis in modern cinema and television as evidence. Known as The Great Patriotic War to Russians and The Last Brother War to neo-Nazis, the meta-narrative that the Allies (the good guys) fought the Axis powers (the bad guys) to save Europe and East Asia is widely accepted today.

This narrative is accurate as the Allies rightfully destroyed both the murderous Nazi and fanatical Imperial Japanese regimes. However, it still represents an interpretation—a fact normally lost on most Westerners. Had the Axis powers won and lived to write their own histories, society would hear something like, "Europe was under siege by hostile foreigners intent on destroying our culture and way of life, and so we rightfully waged a military campaign to defend ourselves." The horrors of the Holocaust either buried or minimized with both Germans and Japanese patting themselves on the back for a job well done defeating tyranny and despotism.

Thus, most label their beliefs about the past as "common sense." When, in reality, they are the product of careful historical interpretation—sculpted by scholars, documentarians, and commentators over decades and centuries. Indeed, the ultimate aspiration of any argument about the past is to stealthy rise to that hallowed ground of "fact." So that when a poli-sci undergrad, or your cousin at Thanksgiving, or some talking head on Fox News confidently proclaims the causes and effects behind an historical event, based on a documentary they watched yesterday—they do so believing they are stating an undisputed fact about reality.

Yes, sometimes the evidence in favor of adopting one particular conclusion is so overwhelming, much like it is for interpreting World War Two, that the job of the historian is redundant. *"I don't need an historian to tell me that the Allies did the right thing defeating Nazi Germany."* Perhaps, but truth is a malleable thing, and present-day understandings may give way to newer, possibly more corrupt ones.

For example, a popular argument about World War Two (no doubt tainted by anti-American bias) is that the Soviet Union "won" the war and the United States barely contributed.[90] At first glance, total casualty numbers on the Eastern Front support the idea. In reality, the United States steadfastly supported the USSR by providing practically all their armored vehicles and air assets. The Army base Fort Wainwright in Alaska was originally Ladd Airfield, where soldiers secretly repainted American planes in Soviet colors to fly them into Russia. Without this level of support, Red Army gains on the Eastern Front would have been impossible. Furthermore, had Operation Overlord not occurred and the

Soviets been allowed to push all the way to the Atlantic Ocean—Paris, Amsterdam, and Barcelona would have all fallen under the Iron Curtain. [91]

Safeguarding the truth against politically motivated contrivances such as "the United States barely contributed to the liberation of Europe" is the ultimate social utility of history. In similar terms, propagandists cosplaying as "historians" have already written and will continue to write accounts hailing the virtuous response of public health officials to COVID-19. They argue that if only those meddlesome and dangerous enemies of "science" had further bent the knee—society could have "saved more lives." Left unaddressed, absurdities like this fester in the collective memory, eventually recreating the conditions for similar debacles to occur again. [92]

Ultimately, any historian must ask themselves "why does my subject matter?" By its very nature, their inquiry lies in the past. Self-help gurus, New Age philosophers and even common sense beseeches us to engage with the present: "Forget the past." "Live in the moment." "Be here now." Legendary golfer Jack Nicklaus said he never thinks about a past shot—ever. So maybe the same should apply to the COVID-19 pandemic. It is true, the horrors of lockdowns, the exposed tyranny of both elites and the mob, the gross miscalculations by "experts," and overall nihilistic haze of that period is over. Should we not just move on and let it go? [93]

Much of the world certainly has. In general, society seems comfortable engaging with detached bewilderment when the subject need arise. Almost like how most look at the exposed motherboard of a computer—vaguely aware of how it works but ultimately lacking any true understanding of its mechanism. Today, citizens feel the aftereffects of the pandemic through increased acceptance of remote work, perhaps they wash their hands more often, or notice residual social distancing markings on their post office floor. Yet, the actual event itself remains a kind of ghostly enigma.

The period is abstract, like every historical period, but it lacks that solidity and reflection major events deserve. Rather, it all feels like a bad dream. Vague memories of being yelled at to "put a mask on!" or hearing local authorities extend lockdown measures "another thirty days to flatten the curve" linger in the collective memory. A recent study published in 2024 found the pandemic even distorted our sense of time itself. Of course, the impact of printing seven trillion dollars to avert total economic collapse is felt everyday through inflation and increased wealth inequality. 94

However, despite its enormous effects, most feel comfortable letting it all go. Citizens rightfully want to get on with their lives, simply feeling grateful the madness is over. So why continue to address it? The same could be asked of the Punic Wars or the Reformation or slavery in the United States. "It's over. Why should I care?" Despite this honest question, the fact is that history infuses every part of our lives. Memories of the past, accurate or not, shape the foundation upon which the present is constructed. Thus, as the COVID-19 pandemic comprises one of the most seminal moments in human history, its lessons must be unraveled in full detail.

So what explains the curious lack of reflection? Perhaps, it is too soon. Perhaps, some do not wish to grapple with the ways in which they behaved, or how something buried deep within them lashed out. Perhaps, the era is simply too painful to reflect upon. Nevertheless, this is the burden of the historian: to dig up the past, present it, and judge it for better or worse.

In *The Gulag Archipelago*, Soviet dissident Alexander Solzhenitsyn embodied this duty. He recounted how Stalin's secret police arrested millions of innocent people and sent them to a network of labor camps. For decades, agents from the People's Commissariat for Internal Affairs, known as the NKVD, approached Russian citizens on the street or in their homes. Agents told citizens they were suspected of a crime and demanded they report to a local police station. These "arrests" were conducted peacefully and cordially so as not to arouse panic in the victim or bystanders. The vast majority of detainees would voluntarily follow the police thinking, *"this must be a simple mistake. I am innocent, of course, and once I speak to the authorities, this will all be cleared up. I will be back home in a few hours."*

Nothing could be further from the truth. Innocence or guilt was irrelevant as Stalin merely needed to instill fear in the Soviet population.

Once separated from their homes and families, victims were forcibly transported to gulags in remote locations around the country where millions would die of disease, starvation, and overwork. The tale is familiar to anyone even vaguely aware of the horrors of early twentieth-century totalitarianism.

However, Solzhenitsyn also described something peculiar and possibly incomprehensible at first glance. A tiny minority of Russians, when the NKVD knocked on their door, either snuck out a window or back door. Ironically, these individuals had now committed the actual crime of fleeing police detainment. One would think the NKVD would begin a manhunt and chase them down, sending the message to others that fleeing would result in arrest.

Instead, the agents ignored them. For it did not matter who was being arrested, only that arrests were being made. Officers simply moved down the hall or across the street to detain someone else. The thought is almost comical, if not for the horrors it entails. But the image of a Soviet policeman merely shrugging his shoulders and casually knocking on the door of another unsuspecting victim is numbing, trite—banal.[95]

The whole terror of Stalin's regime is yet another tally in the long list of episodic madness that has engulfed civilization. History has proven that organized hierarchies backed by violence descend into periodic spasms of disorder. While deliberate abuse by corrupted elites is certainly to blame, their coercive abilities depend on the willing surrender of subjected populations. What if every Soviet citizen had bolted when an NKVD goon knocked on their door? Or shot back even? Solzhenitsyn argued that had Stalin not confiscated firearms then the whole system of mass arrests would have been impossible.

Nonetheless, in the United States, despite their right to bear arms enshrined in the Second Amendment, citizens permitted their own subjugation. American society allowed a panicky fourth estate and lecherous public health establishment to rip away its freedoms in response to a cold virus. The victor of two world wars, a worldwide beacon of individual liberty, and greatest hope for mankind reduced to a horde of masked, terrified subjects confined inside their homes.

Once the tyranny finally ended, it was not because of lockdowns or vaccines or "flattening the curve" or any other nonsense. The virus continues to spread to this day and will continue to do so. Rather, it "ended" because the populace finally said enough! For it is up to the subjugated to decide whether the madness will continue.

***

In Communist Cambodia, dictator Pol Pot sought to erase any history of the country to usher in an agrarian utopia. For Pot and the Khmer Rouge, achieving "real communism" entailed a total reset of Cambodian society. Part of that reset meant eviscerating any memory of the past, any traditions or cultural remnants—in other words, a return to "Year Zero." Books were burned, universities shut down, and even wearing glasses was criminalized, as doing so connoted knowledge and the ability to read. Pot proceeded to depopulate the cities, forcibly removing their inhabitants to the countryside to work in agrarian communes.[96]

Spoiler alert: the agrarian utopia did not materialize. Rather, mass death and suffering the likes of which rival Maoist China and Stalinist Russia consumed the country. Pot executed all members of the "Old Society," meaning lawyers, doctors, teachers, engineers, and intellectuals. A prison was established called S-21 where thousands underwent heinous tortures and experimentation in procedures that make those of Mengele or Japan's Unit 731 seem tame. Over one-fourth of the entire Cambodian population was killed. Year Zero and the erasure of history meant nothing more than a senseless descent into horror and despair. May we never get to Year Zero. May we always remember the past, learn from it, and improve on its iteration.[97]

So, at last, we return to that nether zone of unspoken experience. There has been a curious lack of reflection about the pandemic. Despite its massive convulsions and persisting scars—the era feels almost forgotten. Sure, our fast-paced world and proclivity towards "running to the next fire" is partly to blame. Indeed, "moving on" is one of humanity's greatest strengths. The happiest and most fulfilled are usually those capable of letting go of the past. Humans appreciate history but ultimately transcend it by creating new and better conditions.[98]

Yet, there is something more at play when it comes to "moving on" from COVID. That mask—that flimsy mask—came off. The chaos that simmers underneath was unleashed to reveal something far more disfigured. However, the places we dread are the ones we must travel to the most. As Carl Jung said, "In filth, it will be found."[99] Well, here it is, in the words of another historian: not a piece for the moment, but a possession for all time.[100] A reminder of our capabilities: to build, create, heal, and triumph but also to subjugate and oppress. A reminder that civility, liberty, and sanity are fragile—easily ripped from our lives. In homage to every historian who has ventured into that pit of past experience, and to the field itself, let us vow to never forget its lesson nor leave it to propagandistic hacks. In other words, let us have our reckoning.

THE ROAD TO WUHAN

"Warfare is the Way of deception. When active, appear inactive.
When near; appear far. When far; appear near. When strong, appear
weak."

— *Sun Tzu, The Art of War*

The Europeans came in waves, ripping open China like a flesh
wound. The French took Guangzhou; the British grabbed Hong Kong.
Germany, the Shandong region, and Russia, Manchuria. Shanghai was
carved up and Peking occupied. At the turn of the twentieth century, after
two Opium Wars and a near endless stream of unequal treaties, the
Middle Kingdom was a client state.

Ironically, only a few hundred years earlier, a fragmented Europe
stagnated amidst feudalism and cultural decadence whilst Imperial China
enjoyed a golden era of wealth and art. With far greater manpower and
naval assets, the Chinese could have charted the globe, carving out their
own empire in the Americas and Pacific. Though unlike their European
counterparts, imperialistic conquest is antithetical to Chinese culture. The
word for "China" in Mandarin is *zhongguo* which literally translates to
"middle country." Meaning for millennia, the Chinese thought of
themselves as occupying the center of the universe. Land, gold or empire
is of little value at the heart of all things.

So, despite greatly surpassing the West in economic and cultural
development; instead of expanding—they retrenched. The Qing Dynasty
barred every one of its ports to foreign traders except Canton (and even
that came with extensive restrictions). They rejected imports of
technology like steam, rifled firearms, and telegraphy. They banned the
consumption of Western literature or art and shut off diplomatic relations
with other countries—as what could the "barbarian" outsiders possibly
offer All Under Heaven.[1] Nonetheless, the steamroller of history churned
forward and by the mid-nineteenth century, the erstwhile inferior
Europeans eclipsed China in almost every metric. Despite heralding the

invention of gunpowder, China was now deeply disadvantaged militarily having eschewed advances in firearms, artillery, and warships. At the same time, the country's vast quantities of tea, silk, and other precious goods set them on a crash course with European powers desperate for new markets.

The result is well known. Iron-hulled, steam-powered vessels raising the Union Jack poured forth across the Chinese coastline. Imperial soldiers bearing swords, muskets, and smoothbore cannons stood little chance against three centuries of technological advancement. Indeed, isolationism may have preserved Chinese culture (maybe even made it superior) but that only goes so far when looking down the barrel of a British artillery piece. Thus, the Crown got what it wanted, opening numerous ports to trade and securing Hong Kong.

Yet the British—always grasping for more—attacked again twenty years later with the same result. This time unsealing all of China with virtually every restriction on foreign trade removed. Boatloads of tea, silk and porcelain made their way to Europe to be sold at huge markups. Nevertheless, despite being forced into exporting their own goods, the Chinese wanted little of what the British could offer besides silver. Centuries of conditioned revulsion to foreign influence still dominated the Chinese psyche. Therefore, the resulting trade imbalance decimated silver reserves of British merchants, forcing them to find something outside of China their hosts desired.[2]

Drugs have always influenced history. The introduction of caffeine in Europe helped spawn the Enlightenment and Industrial Revolution. The effect of psychedelics upended U.S. culture in the late 1960s, influencing civil rights and anti-war movements. The Nazis used amphetamines to fuel their rapid, relentless pace of Blitzkrieg by keeping soldiers awake, alert, and aggressive for days without rest. In China's case, highly concentrated extracts from opium poppy plants devastated their society. British manufacturers established large-scale production facilities in India, flooding the Chinese market and addicting millions to the drug. Doing so not only rectified the trade imbalance by preserving European silver reserves but further domesticated the Chinese population. Resisting colonialism while lounging in an opium den is rather difficult.[3]

The humiliating experience inculcated an intense aversion to narcotics within China that still exists today. Even relatively benign drugs are highly criminalized, as painful memories endure of drug-stupefied Chinese subjects powerless to foreigners pillaging the country. Not only that, but a century and half later when production and trafficking into the West of an opioid derivative—fentanyl—boomed in China's criminal underground, its leadership did not exactly leap at opportunities to stop the practice. Understandably, the Chinese have little sympathy for those suffering from the importation of highly addictive substances and resulting incapacitation of their populace.[4]

The overarching era of foreign domination represents a deeply painful and humiliating chapter in Chinese history. A period which lasted more than a century, ending only after the Second World War. Given China's relative dominance vis-à-vis the West only a few hundred years earlier, the experience would resemble Afghanistan or the Congo or Panama crushing the U.S. military and establishing autonomous cities along its coasts and rivers. It is no wonder the Chinese vowed that never again would the Middle Kingdom be made to kneel before a foreigner.

One of dozens of port cities forced open was the small, riverside city of Hankou. Sitting at the mouth of the Yangtze and Han Rivers, the city was ideally positioned to become a major trading center. Outright annexation or colonization was too difficult given complex imperial governance structures and the surrounding hostile population. Yet, British merchants needed free reign in trading with nearby suppliers and circumventing exportation duties. In what would become the standard model for all European outposts operating throughout China, the British established a "concession" in Hankou.

Functioning as a practical mini-colony—but maintaining nominal sovereignty within imperial hands—the concession operated under its own laws and customs. The enclaves often barred Chinese citizens from entering plus established legal systems that disadvantaged locals—an arrangement that effectively equated to foreign occupation. Keeping in the tradition of foisting unequal treaties onto the Chinese, British merchants treated their native counterparts poorly, forcing them to agree to unfair trade agreements. In short, the Chinese had been relegated to second-class citizens within their own borders. Realizing Britain was content to protect its shipping and residency privileges without resorting to armed conflict, other European powers along with Japan moved to establish their own concessions throughout China—including Hankou. By the end of the nineteenth century, France, Russia, Germany and Japan each possessed autonomous zones within the city.[5]

Working within the Hankou concession, a British or Russian or Japanese merchant could gaze across the Yangtze River into the district of Wuchang. One hundred years later, that same vista would feature towering skyscrapers, malls, KFCs, and other hallmarks of China's economic and social modernization. But in the early twentieth century, small farmhouses, tea plantations and wetlands dotted the terrain.

If you were to travel back in time and speak to one of those merchants overlooking Wuchang, you could tell him three things—only two of which would surprise him. The first, that in only a few years, the cities of Hankou and Wuchang will be administratively but also linguistically combined to create Wuhan. Second, you could point to a spot in the wetlands across the river and tell him a laboratory dedicated to the study of infectious diseases will someday exist there. While possibly surprised at the idea of an institution of science being erected in China (as only major European and American capitals housed scientific institutions at the time), the prospect of studying living things at the molecular level comprising the cutting-edge of future infectious disease research would not have astonished him.[6]

This is because that merchant would likely have received some degree of secondary education, meaning a rudimentary understanding of recent advances in science. Some of those perceptions would be thoroughly debunked, most notably the inapplicability of Newtonian physics at the quantum level or at very high speeds. However, their conception of disease transmission closely resembled our own.[7]

While the structure and behavior of viruses was little known at the time, most accepted that microscopic living organisms transmitted from host to host caused illness—known as "germ theory." Science had at last progressed from outdated concepts of pathogens traveling in "bad air" or vapors. Termed "miasmic theory," plague doctors wearing bird-like, beaked masks carried around herbs and spices to "clean" infected air. A practice not too dissimilar in terms of effectiveness nor naivety than people wearing plastic masks during the twenty-first century to combat an aerosolized pathogen.[8]

Nonetheless, by the early 1900s, understandings of disease transmission had made leaps and bounds. Sterilization of medical equipment, handwashing, and other sanitary practices were correctly believed to reduce the transmission of illness. Therefore, the mere presence of such a laboratory in Wuchang would not have shocked a Hankou merchant. What would have come as a shock, for not only being an insult to humanity, but also given their recent experience living through one of the deadliest pandemics in history—the Spanish Flu—would be to inform him that its scientists would successfully attempt to render these "germs" *more* infectious to human beings.[9]

***

Details of the pandemic's origins in Wuhan will likely never be known. Closed, authoritarian societies do not usually conduct free and open investigations into embarrassing and highly destructive historical episodes. Additionally, the Chinese Communist Party (CCP) is dependent on a social contract with its citizens that trades security and prosperity in exchange for censorship of speech and reduced participation in the political system. The word "citizen" is used loosely here as the average Chinese national can hardly be described as holding political rights or privileges in the same way their counterparts in the West do. However, its antonym—"subject"—does not apply so easily as well. The lower and middle classes have found ways to steer their political destiny despite non-existent voting privileges. "Managed stakeholders" or "conditional citizens" is a more appropriate term—or maybe a political science professor at Berkeley or NYU can devise a better one (gaining tenure in the process).

Nonetheless, for purposes of brevity, "citizen" and "subject" will be used interchangeably to describe members of the Chinese population. Considering both Western and Chinese governments locked "citizens" indoors, censored information online, and trampled over their basic freedoms implies our own conception of citizenship needs revision as well.

Nevertheless, Chinese nationals are not the mindless, communist hordes as caricatured by some in the West. Commentators, usually on the right, enjoy depicting a homogenous mass of 1. 4 billion loyal comrades ready, willing and able to serve Marxist-Maoist doctrine. Doing so gives an ambitious talking head or congressperson the opportunity to "talk tough" on China and boost their public image. In reality, Starbucks, McDonald's, Apple, Gucci and all the other hallmarks of the materialistic, bourgeois capitalism Mao and Marx railed against, dominate the Middle Kingdom today.[10]

Moreover, when speaking of the "Chinese," it must be emphasized that the vast majority of regular citizens do not subscribe nor care about the tenets of communist thought. The illiterate farmer in Jiangsu and the Lamborghini-driving, son of steel magnates in Shanghai, equally care about overthrowing the capitalist superstructure—that is not at all. And, frankly, neither does the leadership. Following the failures of collectivist policies in the mid-twentieth century, the CCP abandoned Marxism, embracing a state capitalist economic model.[11]

Nonetheless, economic liberalization did not entail political liberalization as many in the West predicted. Political science theorists assumed increasing wealth would create a wider share of stakeholders, and thus a wider share of people inclined to care about public policy.[12] Even a cursory look at China's past would have dissuaded this view. Chinese history is wracked by internal discord, warring political factions, and social upheaval. For centuries, hundreds of feudal states, kingdoms, and fiefdoms descended the region into bouts of continuous conflict, killing millions. Despite the appearance of control, the CCP is deeply afraid of a return to this *status quo.*[13]

Watching a session of the National People's Congress or glancing at a photograph of the Standing Politburo Committee (the nine most powerful men in the country) gives the impression that China is a homogenous and cohesive body of like-minded individuals. The truth is that the country is deeply divided along ethnic lines, encompassing an enormously diverse body politic bound only loosely by national affiliation. Buddhists in Tibet and Muslim Uyghurs in Xinjiang threaten to upend the patchworked amalgamation by gaining independence. Wealth inequality between inland rural areas and coastal elites risks economic instability. Urban, educated progressives who have traveled to the West and experienced political freedom first-hand threaten increased democratization.[14]

Most importantly, the financial and political elite are almost uniformly Han Chinese—the race of the dynastic class and historical rulers of the country. Official statistics claiming the population is ninety percent Han Chinese are a lie, as the country is much more racially divided. In essence, modern China is a Han ethnostate, with opportunities for political and economic advancement reserved only for that group. Like in any state undergirded by race, the dynamic fosters resentment and revolutionary sentiment within disadvantaged groups.[15]

Put more simply, China is too disparate ethnically and politically to permit a multi-party, democratic system without risking total disunification. A brief experiment with republicanism following the fall of the Qing Dynasty lasted less than three years before warlords reassumed power.[16]

However, neither is totalitarian control possible. The ostensibly absolute rule of the Party has always involved a degree of negotiation with its populace. Mass starvation during the Great Leap Forward forced Mao to end collectivist economic programs in the early 1950s. Roughly twenty years later, Mao was again compelled to abandon plans to reshape Chinese society amidst the chaos and mass murder of the Cultural Revolution (though that madness would only truly end after his death). In the late 1970s, rebellion against collectivist farming practices in Anhui provoked a general uprising across the country.[17]

As a result, Mao's successor, Deng Xiao Ping, was forced to liberalize China's economy and further open to the West. The resulting explosion in wealth and standards of living lifted millions out of poverty, forming the next iteration of China's social contract. Citizens accept reduced political freedom in exchange for massive state-subsidized economic growth and ironclad national security. The system promises to avoid the intermittent chaos of democracy with its dizzying political discourse and uncertain electoral outcomes.[18]

Yet, any deviation in this controlled system—any sprinkle of madness—is deemed unacceptable. After all, why should Chinese citizens tolerate restricted speech, limited political rights, and authoritarian rule unless total social and economic stability is assured. Therefore, the outbreak of a respiratory virus was initially seen as a golden opportunity to showcase the effectiveness of authoritarian regimes in combating modern threats in ways democracies could not.[19]

A democracy (it was believed) could not shut down and forcibly restrain people inside their homes. In other words, a democracy could not fully stop the transmission of an infectious disease like an autocracy could. The virus would be contained in Wuhan, treated, and eliminated, all in a matter of weeks, while the West toiled in sickness and indecision —or so went the plan in Beijing. Like Fauci, Trump, and the entire media-hypochondriatic complex, the Chinese Communist Party failed to account for COVID-19 being aerosolized.[20]

***

The lockdowns implemented in China make the draconian ones in the United States look libertarian. Indeed, those critical of the response in the United States should at least be grateful not to have endured the "Zero-COVID" restrictions of Chinese citizens.

Around the country, public health officials dressed in hazmat suits forcibly constrained citizens inside their homes and apartments. Mandatorily downloaded apps tracked residents via their cellphone, restricting their movements. Provincial governments designated specific zones of inhabitance for each citizen and issued smartphone notifications instructing them to return immediately if they ventured outside.[21] Drones with speakers blasting messages to stay indoors buzzed incessantly above city blocks. Officials violently removed those who did test positive from their homes and transported them to "quarantine facilities." When residents in major metropolises like Shanghai or Guangzhou did work up the courage to venture outside, they often struggled to find basic necessities like food. Those who did resist confinement orders were frequently beaten in public by CCP officials.[22]

Remarkably, some of these measures were applauded by the Western press during the early days of the pandemic. The *Harvard Business Review* wrote glowingly about China's involuntary contact tracing program and recommended countries in the West implement similar initiatives. That paragon of journalistic excellence—*USA Today*—argued evidence from China proved that "aggressively limiting public gatherings and social interactions can help stop transmission of COVID-19," recommending U.S. authorities adopt similar measures. Andy Mok, a fellow at the Center for China and Globalization, a public policy think tank based in Beijing, no doubt completely neutral and under no sway from the CCP whatsoever, beamed about the "systematic, comprehensive and coordinated" Chinese response, which had "flattened the curve."[23]

Michael T. Osterholm, director of the Center for Infectious Disease Research and Policy at the University of Minnesota, proclaimed "I think they [China] did an amazing job of knocking the virus down." Even President Trump opined that "I think China is very, you know, professionally run in the sense that they have everything under control." In maybe the most laughable conjecture of all, the World Health Organization reported that the Chinese government had successfully stopped the transmission of COVID in April 2020, calling their interventions the "only measures that are currently proven to interrupt or minimize" transmission.[24]

If falling head over heels for Chinese propaganda and believing that total cases of a coronavirus in a country with 1.4 billion people miraculously plateaued at eighty thousand that spring, *then they were all completely correct*! In reality, the airborne pathogen continued to voraciously spread through one of the most densely populated countries in the world with the Chinese simply banning testing for the virus that summer.[25] Continuing to test would expose the absurdity of claiming the virus was contained, further destabilizing the nation. Instead of calling the

CCP out for their blatant manipulation, the WHO merely echoed the lie that everything was under control.[26]

Of course, the WHO was deeply compromised by China at the time. This is only natural considering the CCP handpicked its director-general Tedros Adhanom Ghebreyesus to lead the organization. Thanks to steadily rising donations beginning in the early 2000s, the WHO was a practical subsidiary of the Chinese government, and Ghebreyesus, an employee. The exclusion of Taiwan into WHO further speaks to the immense sway China held, and continues to hold, over the world's premier public health authority.[27]

Money buys influence, and the United States is certainly not innocent of paying for control over ostensibly "internationalist" institutions—many of which they unilaterally created after World War Two.[28] The Chinese cannot be faulted for attempting to do the same. Issues arise when media outlets *en masse* treat these organizations more seriously than compromised mouthpieces. Arguing that a particular response is invalid simply because it contradicts a WHO recommendation is a fallacious appeal to authority and insult to the reasoning power of individuals.[29]

Nonetheless, when the entire edifice of gain-of-function research came crashing down with the leak of COVID-19, the Chinese rightfully expected a return on their investment. The WHO upheld their end of the bargain, praising how effectively the CCP responded to the virus and dutifully denying any suggestion it may have leaked from the Wuhan laboratory. Additionally, by the end of January 2020, just as the world was becoming aware of the new disease, Tedros lauded the CCP's "transparency" and "openness to sharing information."[30] In reality, Chinese officials had brutally stifled any mention of the virus transmitting between humans, or that its spread was out of control. Search engine results for COVID-19 were manipulated and whistleblowers silenced. Local statistics were suppressed and communication between individuals monitored to maintain secrecy.[31]

The contradictions between official WHO pronouncements and reality only functions to elicit their total exploitation by the Chinese and does not suggest increased transparency may have stemmed the pandemic. Greater openness may have allowed more time for travelers to avoid Wuhan and slightly slow transmission, but a fundamental argument of this book—borne out by basic facts about microbiology—is that frankly, nothing could have stopped the impending spread. The virus was aerosolized. Aerosolized viruses spread widely. In other words, the virus was uncontrollable whether it originated in Sweden or North Korea.[32]

However, it begs the more interesting question of whether China can even be faulted for their lack of transparency? At the end of World War One, U.S. President Woodrow Wilson and other Western leaders promised the Chinese a role in establishing the new postwar, world order. Not surprisingly, they were denied the opportunity, relegated to client status vis-à-vis the West yet again—this time seeing sections of their country carved out to the Japanese.[33]

Again, why should a country who was bullied for centuries by the liberal, democratic establishment abide by its rules? This is not criticism, but a mere observation. Westerners must acknowledge the deep scars embedded on the Chinese conscious and how their motivations are rooted in centuries of mistreatment. The lesson is that anything is justified in China's rise to predominance and rosy expectations of their assimilation into Western norms are deeply misguided.

***

To fully understand the Chinese response to COVID-19, we must travel back in time to that glorious era of iPods, the U.S.-led coalition invasion of Iraq, *Final Destination 2*, and low-rise jeans. The year is 2003, and the CCP is combatting the outbreak of Severe Acute Respiratory Syndrome (SARS). Like COVID, SARS is caused by a coronavirus. SARS-CoV-1 denotes the virus that causes SARS; and SARS-CoV-2 denotes COVID-19. Also like COVID, the Chinese responded to information regarding the spread of SARS with a typical mix of authoritarian manipulation and disinformation.

First identified in Guangdong, SARS is widely considered to have zoonotically transferred from bats over to humans, probably at a wet market. Sound familiar? Nevertheless, when WHO officials traveled to

Guangdong to assess rumors of an outbreak, Party officials packed SARS patients onto ambulances and drove them around until the inspectors left town. The CCP restricted communication amongst researchers and muffled discussion of the virus in press and government channels.[34]

Nevertheless, on July 5th, 2003, less than eight months after the first reported case, the World Health Organization declared SARS contained and the pandemic over. Official worldwide tallies place the total confirmed case count at 8,096 with 774 dead—having spread to thirty different countries.[35]

Those numbers are almost certainly incorrect and drastically understated. SARS-CoV-1 derives from the same genus as SARS-CoV-2, meaning they are highly similar in genetic and structural composition. Both are spike-proteined and attach to the same cell receptors when attacking hosts.[36] Yet, COVID is credited with infecting over 88,000 times more individuals. Of course, the total COVID-19 case count is dramatically inflated (for reasons which will be addressed later), but supposing the number is only overcounted by one standard deviation, it would still dwarf the total SARS number by over 81,000 times! What explains this?

There are three possibilities:

#1 SARS has a transmissibility rate of 80,000 times lower that of COVID-19 despite sharing the same genus group.

#2 Health authorities in *thirty* countries pulled off the most phenomenal job in the history of public health to contain a respiratory virus in less than eight months.

#3 Something else is responsible.

Let us assess possibility #1—that SARS is 80,000 times less infectious than COVID. To do so, we must examine its Basic Reproduction Value (R). This metric estimates how many people, on average, an infected individual will spread the virus to. The R of SARS is somewhere between 2-4. For comparison, the original strain of COVID-19 exhibited an R of 3-4.[37]

Furthermore, anecdotal evidence suggests SARS was highly infectious. A hospitalized fishmonger in Guangzhou spread the disease to over 105 nurses and doctors. In Vietnam, one patient infected over twenty medical staff. In Hong Kong, one hotel guest is credited with infecting twenty-three other guests. Clearly, the disease was markedly infectious and likely comparable in transmissibility to that of COVID-19.[38]

By the summer of 2003, the virus was so contagious that major health authorities doubted SARS could ever be contained—let alone eradicated. In June of 2003, Jim Hughes of the U.S. Centers for Disease Control and Prevention predicted SARS would remain a permanent fixture in East Asia. Even top Chinese officials were signaling alarm, with Zhong Nanshan, director of the Guangzhou Institute of Respiratory Diseases, publicly describing SARS as "not under control."[39] That same month, SARS had spread to India, Canada, and the United States along with dozens of other countries. The FIFA Women's World Cup was moved from China and officials in Toronto ordered thousands of residents to quarantine in place.[40]

Taken together, it is safe to conclude that SARS is not tens of thousands of degrees less infectious than COVID. Yet, by July of 2003, the WHO miraculously declared the SARS pandemic "contained." In fact, the disease continued to spread even after this declaration, with outbreaks occurring until late 2004. Nonetheless, they declared the virus "eradicated" in 2005.[41]

So, labeling the suggestion that SARS is 80,000 times less contagious than COVID as "unlikely" is charitable at best. Thus, we examine possibility #2: public health authorities across thirty different countries executed the most highly effective pandemic response in world history—an ability they apparently lost seventeen years later.

The argument goes that swift isolation of infected individuals along with strong protective practices like hand washing and mask-wearing reduced the Effective Reproduction Rate (Re) of SARS to zero. Rather than measure the reproductive rate of a disease assuming the absence of immunity and behavioral changes by the population like R does; Re measures the average number of people an infected individual will spread the virus to after immunity develops and public health measures have been implemented. Since the last confirmed case of SARS was reported in late 2004, the Re of SARS is zero. Put differently, in less than two years, the disease was supposedly eradicated. Yet, after only four years since emerging, COVID-19 continues to infect millions of people annually, despite sharing almost identical genetic characteristics with SARS.[42]

So if possibility #2 is incorrect, which it almost certainly is, then the virus continued to spread and infect more people than official statistics maintain. What most likely happened is that SARS started being diagnosed as influenza or just a common cold.[43] After all, coronaviruses are, by definition, cold viruses. In 2003, like every year, members of the population came down sick with a respiratory bug, treated it with over-the-counter medicines and recovered within a few days. In probably not an insignificant number of cases, some had unknowingly contracted SARS-CoV-1. Carriers incorrectly assumed it was either the flu or cold, recovered quickly and went on with their lives. Thus, SARS was never "eradicated," and saying so only perpetuates the myth that aerosolized coronaviruses are containable once introduced into the population.[44]

Finally, possibility #3—that there is more to the story. Real-Time Polymerase Chain Reaction (RT-PCR) testing does not come to mind as one of the most influential technologies of the past quarter century. Nonetheless, few technologies have contributed more, albeit indirectly, to the decimation of human civilization. Before its development in the mid-2000s, testing for infectious diseases entailed a laborious process. Even with the advent of traditional PCR testing in the 1980s, detecting the presence of specific viruses necessitated expensive equipment found only in centralized laboratories manned by experienced professionals.[45]

To best illustrate the process, let us use SARS-CoV-1—the virus that causes SARS—as an example. If you despise technical jargon, the following is only meant to convey the exertion required to test cases before RT-PCR was developed, and you can skip the next few paragraphs. Nevertheless, they describe the process required to diagnose positive cases of SARS during its 2003 outbreak.

First, an individual exhibits flu-like symptoms and reports recent exposure to confirmed SARS cases. A medical professional suspects the patient may have contracted the virus and orders a test. The individual is either physically transported to a Biosafety Level 2 or higher laboratory to provide a sample—or submits one remotely. The traditional PCR process begins when a clinical microbiologist examines the sample through a light or electron microscope. The pathogen is then isolated and its genetic material extracted. Assuming the genome of SARS has already been sequenced (which happened in April 2003), the sample's genetic material is mixed with that of SARS in an extremely high-powered oven, known as a "thermocycler."

Without further getting bogged down in technical details, in essence, the oven repeatedly heats and cools the two genetic strands. If SARS is present, then the sample's DNA will be doubled over and over again in a process known as "amplification." However, this only becomes evident after the sample is removed from the thermocycler and infused with an electrically conductive gel. Once electrified, the negatively charged DNA molecules migrate across the gel to a positively charged terminal, coalescing into more easily identifiable clumps—in a process known as "gel electrophoresis." A microbiologist trained in the process then examines the sample to identify if its DNA had been amplified. If so, the patient has tested positive for SARS-CoV-1.

This is all to say that in 2003 identifying the presence of a respiratory virus like SARS, which is highly similar in makeup to COVID, was an extensive and time-consuming process with only one sample examined at a time.[46]

The exertion required becomes all the more stark when compared to how its successor technology, RT-PCR, operates. In short, RT-PCR did away with the necessity of gel electrophoresis, sifting any amplified genetic material *during* the heating and cooling process. The fusion of both tasks enabled the entire operation to be condensed into smaller machines capable of handling dozens of samples at once.[47]

Fast forward to 2020, and now suspected cases, even asymptomatic and perfectly healthy ones—or simply those curious about their infection status—could drive or walk to a makeshift point-of-care facility, provide a nasal swab, and discover within thirty minutes whether they had contracted the virus. Thus, saying "100 people tested positive for COVID" means something very different than "100 people tested positive for SARS." The latter entailing an enormously larger expenditure of time, energy, and financial resource.[48]

Had RT-PCR been available in 2003, total SARS cases would have undoubtedly been scores higher than the official count. Media professionals could have gone into overdrive stoking fears about an East Asian respiratory virus "killing millions," probably blaming Bush or Cheney for the resulting American epidemic. Lack of smartphones and a slower information environment hampering their hysterics to an extent, but a higher case count undoubtedly results in a very different 2003-04 period. Maybe we would be living in a "post-SARS" world and this book would chronicle *that* pandemic. Maybe a few giddy, public health officials at the height of their careers in the early 2000s, could have assumed total power and authority to restrict the lives of millions of Americans. Alas, they needed only wait a few more years.[49]

This is not to advocate for curtailing the advancement of medical technology like RT-PCR or any other useful tool for that matter. What it does illustrate are the potentially destructive effects of those advancements, and how society interprets their insights matters just as much, if not more, than their mere development.

Regardless, SARS sounded alarm bells in Beijing that something like an infectious pathogen could upend their carefully curated social order. Infectious diseases, as those in the West viscerally discovered in 2020, invoke tremendous psychological responses in certain segments of the population. Reports of spreading viruses trigger intense survival instincts inside these groups which override their capacities for logic and rationality. *"The virus has a 99.9% survival rate for those aged below seventy years old? I don't care, ban everyone inside my state from going outside!"* Furthermore, failure to stem viral transmission is perceived as an indictment on the ruling class and violation of its social contract.[50]

Much of the mania, understandably, emerges in mothers as survival of their children supersedes any feeling of national affiliation or fealty towards a given regime or political system. If a virus primarily targeting the young ever emerges in the near future—satisfying the wishes of Sam Harris—you can count on the complete obliteration of any semblance of freedom, far exceeding that exhibited during COVID. Look only to calls to repeal the Second Amendment in response to school shootings as evidence.[51]

Nevertheless, the psychological effects of pandemics are all the more pronounced under authoritarian regimes. In the case of China, its residents—excluding coastal elites able to travel abroad—are born into, grow up in, and live inside a controlled environment where the Party is deemed the only political entity capable of ruling the country. The Party is not merely the best option, but the only option.

Conditions may deteriorate due to unforeseen circumstances like an earthquake or hurricane, but the overall expectation is that living under CCP rule entails total safety and steady rises in standard of living. Criticism is permitted but limited to petitionary requests for highly specific changes to policy like forcing a factory to stop polluting or increased compensation for land seizures. Suggestions that the Party itself may be flawed, or even more egregiously, that it be replaced—as what began happening after two years of Zero-COVID—is unacceptable.[52]

To combat calls for real political change, the CCP inculcates an idea within its population that citizens live in the best of all worlds, and that the Party will solve any and all challenges. News from the West of political polarization, racial strife, and mass shootings further instills the notion that democracy entails instability, with authoritarianism representing the only safeguard against it.

As Westerners, it is nearly impossible to step into the shoes of an average Chinese citizen, empathizing with the notion that a one-party system is not only ideal, but self-evident. The prospect of the Republican or Democratic Party occupying the Presidency along with every seat in Congress, plus guaranteed victories in every election is so antithetical to American values that its mere suggestion is comical. That being said, the United States discarded basic constitutional rights because a few Ph. Ds and neurotics said so. The line between subjugation and freedom is thin, indeed.

Nonetheless, in anti-democratic states like China, the spread of infectious disease shatters the impression that authoritarian rule is worth its tradeoffs. Therefore, a pandemic ranks only behind defeat in war as the greatest threat to one-party rule.[53]

Hence, by the spring of 2003, SARS had erupted into a near existential crisis for the CCP. Led by President Hu Jintao, the Party's efforts to contain the disease whilst censoring reports of its spread was failing. As they again discovered in 2020, coronaviruses are aerosolized and impossible to contain completely. SARS cases rapidly spread outside of Guangdong, even creeping into the ranks of the Party itself.[54] Social media apps like WeChat and Weibo percolated information about coronavirus transmission a dozen and a half years later; but in 2003, text messaging and internet chat rooms sufficed. In fact, the requisite surveillance infrastructure to monitor and control telecommunications had yet to be developed, allowing Chinese citizens relative free reign in sharing information about SARS.[55]

Furthermore, while hard to believe now, the Chinese Communist Party of the early 2000s sought liberalization and gradual expansion of political freedom. A new generation of leaders, many educated in the West and scarred by painful memories of the 1989 Tiananmen Square protests, envisioned a nation more closely aligned with democratic norms.[56]

With the advantage of hindsight, skeptics argue this was done to deceive the West into permitting China's accession into internationalist organizations like the World Trade Organization—only to have the CCP resort to hostile nativism following their admittance. Maybe so, but more likely, the SARS fiasco dissuaded Party officials from continuing reforms, paving the way for its authoritarian shift under Xi Jinping. Nevertheless, the brief window of freedom at the start of the millennium allowed regular citizens an opportunity to freely exchange information regarding SARS. By early 2003, text messages and e-mails swirled through the Chinese cyberspace, transmitting news of overwhelmed hospitals and lying officials faster than the disease itself. By March, over forty percent of urban residents had heard about the virus through unofficial channels.[57]

Living under authoritarian rule entails some degree of acceptance that you are being lied to. When it comes to infectious diseases, however, even conditioned populations tend to have little tolerance for manipulation. Chinese citizens are not oblivious drones and have demonstrated an adept ability to detect when their government is lying to them. Therefore, in only a few months, rumors and information about SARS morphed into an indictment on the regime itself.

In an emergency meeting of the Standing Politburo Committee in February, Premier Wen Jiabao declared that "the health and security of the people, overall state of reform, development, stability, China's national interest and international image are at stake." The Politburo declared any report of SARS infection a state secret and its publication without official assent from the Ministry of Health, a criminalized offense.[58]

Not surprisingly, controlling information about infectious pathogens is rooted in the Party's very being. After Mao assumed power in 1949, he banned discussion about epidemics as it called into question the Communists' ability to govern. Yet, what may have worked to censor information in the mid-twentieth century was sorely outdated by the digital age. In the first quarter of 2003, Chinese subjects sent over 26.5 billion text messages amongst themselves, many of which discussed the virus. "Foreign reports: SARS has reached Beijing," read one. "Foreign reports: Two dead in Beijing from SARS," read another. Therefore, the Party-sanctioned news blackout only exasperated panic rippling online about the epidemic.[59]

President Hu wanted to allow provincial governments and media outlets to report freely regarding its spread. Being in office for only a few weeks, Hu sensed an opportune time to distinguish his regime from the opaque and censorious style of his predecessors. Yet, the previous president, Jiang Zemin, who still held significant sway over the Party as head of the military, opposed the idea—arguing free reporting threatened the regime's stability. A regional epidemic had evolved into a political dogfight at the highest echelons of Communist leadership.[60]

The two leaders' disparate upbringings shaped their response to the epidemic. Hu grew up relatively poor in the agricultural hub of Taizhou. Witnessing the horrors of the Great Leap Forward first-hand, he grew skeptical of Marxist doctrine and collectivization. Later, during the Cultural Revolution, Hu's father—a tea shop owner—was denounced as a capitalist sympathizer and imprisoned. Publicly tortured in one of Mao's infamous "struggle sessions" he never recovered physically, dying only a few years later. These experiences inculcated an intense desire within Hu to see China transition away from its authoritarian impulses.[61]

Whereas the upheavals of Maoist Communism barely affected Jiang. Raised in the cultural and artistic center of Yangzhou, he was born into an intellectual and elite family. Surrounded by "books, music, art, and political discussion," Jiang grew up privileged and steeped in Marxist thought.[62] He joined the Communist Party in 1947 and quickly rose through its ranks.

Later, as party secretary for Shanghai, a highly prestigious position within the CCP, Jiang was dispatched to address student protesters at Tiananmen. Jiang played a double game, praising protestors and sympathizing with pro-democracy groups, but supporting more authoritarian tactics behind closed doors. On the same day he spoke encouragingly to the demonstrators, he sent a telegram to Party officials supporting their declaration of martial law.[63]

On the night of June 3rd, 1989, two hundred thousand People's Liberation Army soldiers entered Beijing. As Type 59 battle tanks and Armored Personnel Carriers rolled through Tiananmen Square, thousands of protestors were crushed to death or shot by PLA infantrymen. No definitive death count is known but a clandestine telegram from the British ambassador at the time, placed the death count at over ten thousand.[64] Jiang emerged from the bloodbath the Party's most trusted deputy, with Deng Xiao Ping appointing him general secretary and president shortly thereafter.

After ten years at the helm, Jiang deemed it prudent to conduct the Party's first ever peaceful transition of power in 2003. Hu Jintao, widely known for being a passive technocrat, was hand-picked by Jiang, believing he would be easily manipulable. Thus, his relinquishment of the presidency to Hu functioned more as a symbolic gesture rather than a true transfer of control. As a result, Jiang retained significant influence within the Party and his political faction continued to dominate important positions at the Politburo level.[65]

As the country's leadership deliberated how to respond to SARS, Jiang favored silencing any information regarding its spread. The overwhelming lesson from the Tiananmen crisis for Communist die-hards like Jiang was that an unfettered press risked escalating situations beyond Party control. Combatting an infectious disease required tight restrictions over information flows and media coverage. However, the proliferation of text messages and e-mails stymied any such attempt. The message "there is a fatal flu in Guangzhou" was sent over 126 million times by Chinese citizens over the course of three days in February.

The old Maoist framework to restore order was failing, yet the crackdown accelerated. The Health Ministry instructed major hospitals not to report SARS case counts to the media. The Propaganda Ministry forbid domestic press outlets from reporting the World Health Organization's first global warning about the virus on March 15th. The government even continued to deny the virus was contagious until later that month.[66]

Suspecting foul play, doctors from the World Health Organization arrived in Beijing. In response, Party officials packed SARS patients onto ambulances and drove them around the city until the inspectors left. Having not received copious amounts of "donations" yet, WHO officials were not so easily fooled. Its director-general at the time publicly chastised Beijing's unwillingness to cooperate and share information— issuing a travel warning for China on April 2nd.[67]

Unphased, Health Minister Zhang Wenkang, a close ally of Jiang, reported that China was "safe" and that "SARS had been placed under effective control" at a news conference. Zhang blatantly lied, telling the world that Beijing had only twelve cases. Jiang Yanyong, a 72-year-old retired military surgeon, who had visited local Beijing hospitals, became enraged watching the news conference. On April 4th, he sent an e-mail to

Time magazine reporting that over one hundred patients had been hospitalized in Beijing alone—a huge risk for which he was later arrested for. *Time* published the e-mail on its website, which was quickly forwarded throughout China.[68]

As a result, WHO doctors extended their trip in Beijing and international pressure on the CCP to be more transparent intensified. Yet, Jiang and his faction continued to counsel censorship despite reports of the virus spreading into the country's interior. With the international community demanding China take action, the issue had metastasized into a full-blown crisis.[69]

Sensing the situation had reached a boiling point, Hu took action. On April 17th, he convened an extraordinary session of the Politburo, ordering officials to stop lying about the virus and censoring information. He fired Zhang Wenkang and the mayor of Beijing for incompetence and started holding officials accountable for the disease's spread. The next day, state-run media outlets began publishing accurate numbers of SARS cases. By May 7th, eighteen thousand people had been quarantined in Beijing. Moreover, city officials in Guangdong, the epicenter of the outbreak, mobilized over eighty million citizens to clean houses and streets.[70]

Conventional wisdom posits that China's strong, albeit belated, response to SARS successfully suppressed the disease by summertime— permitting the World Health Organization to declare the virus contained in July. The vast mobilization of health workers and mitigation efforts by Chinese officials are credited with achieving this "major public health victory." However, the virus had already spread to *twenty-nine* other nations by the time China decided to act. While some countries like Canada quarantined certain segments of their population, the vast majority of infected countries did nothing to stop its spread.[71]

So what explains the miraculous end to SARS that summer with only 8,097 individuals infected worldwide? Furthermore, if accepted, then how did the global health community seemingly lose its ability to contain coronaviruses only seventeen years later? As explained above, the relative simplicity of testing capabilities in the early 2000s compared to advancements in the late 2010s—PCR to RT-PCR—is largely responsible. However, the outright curtailment of testing altogether is equally to blame.

By mid-May 2003, it was clear that SARS did not represent a major threat to humanity with its symptoms mimicking those of flu and the common cold.[72] Once public health authorities grasped this, they dramatically reduced testing for the virus. WHO and the United States Centers for Disease Control and Prevention both recommended "judicious" testing for SARS-CoV-1 as to "prevent the unnecessary anxiety that would accompany a false positive test in a low-risk situation." This point cannot be emphasized enough. The World Health Organization and U.S. CDC themselves recommended that testing be reduced to limit panic within society. If only similar logic had prevailed in 2020.[73]

Nonetheless, both the Chinese Communist Party and WHO eagerly sought an off-ramp to the first global pandemic of the twenty-first century. Armed with the recognition that SARS symptoms "resembled those of the common cold or garden-variety flu, [and] frequently escaped diagnosis," the decision was made to simply limit testing.[74]

Ironically, while the deceptive practices of WHO and the CCP are disdainful—in many respects, they reacted brilliantly to SARS. Allowing populations to develop herd immunity whilst treating the pathogen as another cold virus was the best of all solutions. Their actions explain why the pandemic did not metastasize into a ridiculous dystopia as life did from 2020 to 2022.

Indeed, it served both the interests of WHO officials and the CCP to prematurely declare SARS contained. The CCP wanted headlines reporting new cases out of the press and a return to their carefully calibrated social order—further bolstering their credentials as authoritarian stewards of Chinese society. The World Health Organization equally sought to establish itself as the world's preeminent defender against pandemics. If aerosolized pathogens are "uncontainable," national governments would inevitably question why billions in dues were being sent to the organization. The fiction of coronaviruses being controllable had to be maintained.[75]

In hindsight, the WHO declaration of SARS containment in July 2003 is preposterous. Even by May 2004, outbreaks in China continued to occur. For any real journalists out there, researching the sudden plateau of SARS cases would be a highly worthwhile inquiry. Nonetheless, ending testing rightfully removed SARS from newspaper headlines, mitigated panic, and returned the world to normalcy. No doubt, the World Health Organization and CCP sought a similar result in February 2020 in declaring COVID-19 contained.[76]

Born too early to witness the Chinese and WHO manipulate statistics of a pandemic. Born too late to witness the Chinese and WHO manipulate statistics of a pandemic. Born just in time!

***

Just as text messaging rendered Jiang's plan to censor information about SARS obsolete, RT-PCR rendered the same for Xi Jinping's plan to "vanquish" COVID. While perceived as a Hu Jintao-esque, consensus-driven leader prior to assuming the presidency, Xi co-opted the levers of power within China to assume levels of centralized authority not seen since Mao. Thus, he viewed the outbreak of COVID as a prime opportunity to showcase the effectiveness of authoritarian rule. Whether he truly believed SARS had been "contained" in 2003 or not, he sought a similar result in early 2020 by launching the zero-COVID initiatives.

However, the increased testing capabilities of RT-PCR nullified any containment strategy, no matter how draconian or restrictive. Xi made the decision to practically stop testing in March at 80,000 cases as a result. Even today, the total number of cases in China has been frozen at 503,302 since January 11th, 2023. A laughable figure in a country with 1.4 billion people.[77]

Another piece of conventional wisdom following SARS was that China had "learned its lesson" not to censor information about viral epidemics. The BBC reported the experience had convinced the CCP to "work with other countries" and to not "cover it up." A since retracted

2019 article by researchers at Shandong University surmised that "China is better prepared than ever for epidemics" and recognized the necessity of "honesty and transparency" in combatting future outbreaks.[78]

So much for that. On December 30th, 2019, Dr. Li Wenliang, a Wuhan ophthalmologist, warned colleagues in a private WeChat group about a new emerging SARS-like virus in Wuhan. Days later, he was summoned by police authorities, forced to sign a document accusing him of "spreading rumors," and made to recant. Human-to-human transmission was officially denied until January 20th, 2020, despite mounting evidence to the contrary. Posts and hashtags about the virus were deleted or blocked, independent journalists disappeared or detained, and state media continued to claim conditions were under control.[79]

Rather, the lessons the CCP learned from SARS were ones which would precipitate the disaster to come. First, they realized that controlling information flows internationally was critical to controlling China's global image, thus necessitating increased donations to the World Health Organization. A compliant and submissive WHO would be essential to negotiating the next outbreak. More importantly, billions were allocated to studying respiratory viruses and developing vaccines in a vain attempt to control future epidemics.[80]

The CCP designated the Wuhan Institute of Virology as the country's premier laboratory for infectious disease research, prioritizing gain-of-function studies on coronaviruses. Researchers at WIV collected thousands of bat stool and blood samples in an attempt to compile a database of all extant coronaviruses. The logic being that if one did emerge in humans, its genome would already be sequenced enabling faster vaccine development. More thoroughly addressed in Chapter Three, studies at WIV funded by the National Institutes of Health sowed the seeds for the impending pandemic—as yes, *COVID leaked from the lab.*
[81]

The fact that the origins of SARS-CoV-2 were ever in question, and continue to be in question, speaks to the lunacy of contemporary life. Major news publications still operate under the delusion that one of three laboratories in the world experimenting on making coronaviruses transmissible to humans had absolutely nothing to do with the outbreak of coronavirus in that same city in which the laboratory operated. The mental gymnastics required to make this claim defies belief. Moreover, refutations of the "lab-leak theory" from supposedly respected scholars in scientific publications is equally abhorrent.[82]

Here, we make a statement that is so obvious it should not even require repetition: the novel coronavirus escaped from the novel coronavirus laboratory experimenting on novel coronaviruses. Is the evidence to assert this claim merely circumstantial? Yes. Is there definitive proof that enables us to make this claim with 100% certainty? No. However, not accepting this opens up an entire epistemological firestorm untethered from rational thinking.

Can we prove it was not a coincidence that the last time you touched fire it hurt, and that it will hurt the next time you touch it? Can we prove the universe did not pop into existence five minutes ago and that you have fake memories? Can we prove the existence of ancient civilizations like Egypt and Rome despite no one living having seen them first-hand? Denying the origins of COVID from the Wuhan Institute of Virology based on appeals to circumstantiality refutes the nature of truth itself.

Another mind-boggling development was the labeling of those who made the obvious connection that the virus originated from the Wuhan lab as "racist" and "conspiracy theorists." Indeed, American "journalists" characterized the claim as some kind of racial aspersion against the Chinese. Those paragons of journalistic integrity at *The New York Times* called the idea that the coronavirus escaped from the coronavirus lab a "fringe conspiracy theory."

Totally neutral and unbiased journalists at *Slate* labeled asking questions about the lab "good old-fashioned racism." Apparently, good old-fashioned common sense was out of style. Let us not forget these same publications advocate for affirmative action in college admissions, a policy that greatly disadvantages Asian Americans.[83]

Nevertheless, while openness about an infectious disease may have been possible had the virus emerged organically, addressing one that leaked from your country's own laboratory is another matter altogether. Obviously, Xi and the Standing Politburo Committee were informed at some point that the virus leaked from the Wuhan Institute of Virology. Being one of the most closely guarded secrets in the CCP, there is no record or history of this.[84]

However, just as it does not require a colossal leap of logic to determine the virus leaked from WIV, it requires an equally short leap to conclude Party leadership became aware at some point—likely early on. While CCP leaders may have wished to be more transparent about epidemics following SARS, the knowledge that COVID stemmed from their own decision-making apparatus changed the calculus entirely.

There are those who wish to ascribe malice to the CCP in either deliberately releasing the virus or allowing it to escape. The point is unprovable and there is no scenario where enough evidence can be gathered to draw a conclusion either way. More importantly, to return the state of public discourse to something resembling sanity, we must acknowledge that the virus, at minimum, originated from the Wuhan Institute of Virology. Not doing so borders on mass psychosis. Of course, the virus leaked—accidentally or not—from the laboratory.[85]

Furthermore, if done deliberately, the ploy completely backfired. Not only did the pandemic greatly reduce China's international standing, but its leadership faced the most significant challenge to its authority since Tiananmen as a result. After two years of draconian Zero-COVID restrictions, widespread protests erupted across the country. In Xinjiang, already a hotbed for anti-CCP sentiment, at least ten people died in an apartment fire after officials refused to allow residents to evacuate due to lockdown policies. Some of the fiercest protests swept the region in response. By November 2022, demonstrations demanding lockdown restrictions be lifted had spread to nearly every single major Chinese metropolis. Protesters chanted "We want freedom, no more lockdown!" and "Down with Xi Jinping and the Chinese Communist Party!"[86]

Only after it became abundantly clear that sustaining attempts to eradicate an aerosolized pathogen was precipitating calls for regime change did the Party liberate its population. On December 7th, the Standing Politburo Committee abruptly ended lockdowns, local travel restrictions and mass testing. Thus, similar to how lockdowns in the United States only ended after continuing to do so became politically and socially untenable—not because the virus had actually been eradicated—China was forced to do the same.[87]

Likewise, just as the West has tried to "forget" the entire debacle, China has attempted something similar. Though unlike the West, the CCP employs a vast censorship network to accomplish the task. Protests grappling with past injustices during zero-COVID are quickly disbanded. Art exhibits reckoning with the era have been shut down.[88] Mentioning "COVID" or the "pandemic" on Chinese social media is censored with citizens forced to use words like "face mask era" to obviate bans. Yet, like most Americans, regular Chinese citizens wish to let it all go too. *The New York Times* quoted a Shanghai restaurant owner, Fu Aiyang, saying "even thinking about it is painful. Let's not talk about it." Forced to smuggle rice from her restaurant because she lacked food at home due to lockdown restrictions, she understandably wants to move on.[89]

Nevertheless, Chinese citizens have reported lasting mental health struggles. Hou Feng, a 31-year-old programmer from Shanghai, witnessed a screaming neighbor be dragged from their home after testing positive for COVID. Nightmares continue to haunt him of men in white hazmat suits breaking down his door and hauling him off to a quarantine center. No matter how much governments or even regular citizens may want to eviscerate memories of the pandemic—in China and the United States—they will endure.[90]

The ultimate conclusion from China's experience with COVID is a sad one. While Western regimes can be faulted for their tyrannical reaction given ostensible commitments to freedom and civil liberties, the same cannot be said of Communist China. In a sense, their response was inevitable.

Reacting sensibly, meaning allowing the virus to transmit through its population and develop herd immunity was always an impossible task for the CCP. The nature of authoritarian rule necessitated a brutal crackdown. Yet doing so exposed the inherent weakness of all anti-democratic regimes. The ancient Chinese proverb *zhuo jin jian zhou* translates to "when you pull at your clothes, your elbows stick out." In administering their patchworked and disjointed amalgamation of ethnic groups under a one-party system, the Party must inevitably address crises in draconian and brutal ways.

Thus, the CCP are neither cold-blooded Marxists intent on world domination nor accommodating, democratic-aspiring, future friends of the West. They are rational actors, rooted in their historical experience. From the arrival of British warships and hostile importation of addictive drugs, to the SARS epidemic of 2003—the Chinese Communist Party responds rationally based on past experience.

Furthermore, while Chinese culture deemphasizes militarized combat, it values deception and manipulation to achieve political objectives. In the most influential treatise on strategic thinking in China—*The Art of War*—Sun Tzu wrote, "Warfare is the way of deception. Thus although you are capable, display incapability. When committed to employing your forces, feign inactivity. When strong, appear weak. When close, appear far." Expecting anything but this kind of behavior from Chinese leadership is naïve, and Western society must approach its relations with the country cautiously.[91]

Yet, both sides can learn to co-exist peacefully, and even learn from one another. The country's magnificent public transportation network, extensive and high-quality infrastructure, and general transformation from East Asian backwater to global, economic superpower speaks to a potent mixture of command economics, massive populations, and strong genetic traits.

Contrary to jingoist proclamations by American politicians, regular Chinese citizens are more alike than different to their Western counterparts. Walking the streets of Shanghai or Beijing is at times no different than doing so in an American capital with Starbucks, Louis Vuitton and Apple storefronts saturating both environments. Their embrace of Western materialism speaks to a close relationship between the two powers. Similarly, their attempt to forget the ravages of the pandemic evidence similarities between our cultures. Lastly, the Chinese ethos of anti-imperialism portends a positive relationship with the United States. War is not inevitable. May the two great civilizations live in peace and prosperity.

THE GOOD DOCTOR

"The NIH has not ever and does not now fund gain-of-function
research in the Wuhan Institute of Virology."
— *Dr. Anthony Fauci, sworn testimony to the United States Senate,
May 11, 2021*

The President of Egypt asked the Prime Minister of Israel to end
raids on Lebanon. President Ronald Reagan reneged on his promise to cut
taxes for businesses. Secretary of State Alexander Haig hinted the United
States was considering removing China from its list of Communist
countries. These were the major headlines from America's newspapers on
June 5th, 1981.[1]

Though the most consequential item of news, one which would
reverberate for decades, did not appear in any publication that day. Buried
within a weekly report on contagious pathogens published by the Centers
for Disease Control and Prevention, researchers described how clinicians
at three different hospitals treated five young men for a rare form of
pneumonia. Two had died with the other three in critical condition.
Known as *Pneumocystis* pneumonia, the disease had typically only been
observed in patients with severely weakened immune systems. Yet, the
findings were largely ignored by the medical community, overshadowed
by more pressing concerns like Legionnaires' Disease and Ebola.

However, it caught the eye of America's soon-to-be most famous
doctor. Sitting in his office on a pristine summer morning at the National
Institutes of Health campus in Bethesda, Maryland, Dr. Anthony Fauci
picked up the CDC report. Initially surprised that such a rare form of
pneumonia had been observed in multiple cases across scattered
locations, Fauci noticed something even more perplexing: all five of the
patients were gay.[2]

Anthony Stephen Fauci, born the same day Luftwaffe bombers
rained explosives down on British civilians, was the son of first-
generation Italian Americans. Amidst games of stickball and occasional

scraps on the streets of Brooklyn, Fauci excelled academically from an early age—demonstrating a propensity to overachieve which would follow him throughout his life. As an undergrad at Holy Cross, Fauci double majored in pre-med and the classics—though he probably could have gone for another reading of Diogenes considering his actions sixty years later.

In any case, Fauci went on to Cornell Medical School, where not surprisingly, he graduated number one in his class. A prestigious internship and residency followed, after which a whole cornucopia of lucrative jobs at hospitals or in private practice lay open before him. Nevertheless, a proclivity toward public service and burgeoning interest in contagious illness led Fauci to join the NIH's National Institute of Allergy and Infectious Diseases (NIAID). Though, after nearly a decade of working in the narrow field of autoimmune disorders, Fauci felt "unchallenged" and yearned for something more. He would get just that.[3]

Human Immunodeficiency Virus (HIV) is almost diabolical in its makeup. Most families of viruses—like coronaviruses—attach onto host cells, clinging for life and sustenance. Alerted to the presence of a foreign body, our immune system kicks into overdrive, dispatching T cells to kill infected cells and free-floating pathogens before they can spread further. In the vast majority of cases, this layered defense works brilliantly to not only kill foreign invaders but remember their makeup and vulnerabilities —so if the same or similar toxin returns, the body can more efficiently respond. Most of the time viruses infiltrate our bodies without us even knowing it, having been quickly neutralized by built-up immunity.

HIV, on the other hand, transcribes its own DNA onto that of its host's DNA, intertwining the two—permanently. Meaning every time an infected patient's genome reproduces new cells, it also reproduces the virus. This is why HIV is incurable and patients struggle with its effects for life. Not only that, but the virus has a long incubation period, meaning a significant amount of time can pass before infected individuals begin exhibiting symptoms, leading to increased transmission. Finally, and most importantly, HIV virions are best suited to attach onto one particular type of host: immune system cells. Meaning the body's only defense against the virus is the one most compromised.

The result is Acquired Immunodeficiency Syndrome (AIDS) which wrecks the body's natural defenses, leading to increased susceptibility to a whole host of other diseases like pneumonia and tuberculosis. It is safe to say that if HIV were airborne, like COVID-19, we would face a near existential threat to humanity. Instead, HIV only transmits through bodily fluids like blood and semen, meaning sexual transmission is the primary driver of infection.[4]

Yet, on that crisp summer morning in 1981, none of this was known and would remain undiscovered for years. Nevertheless, a forty-year-old Fauci, armed with two decades of immunology experience, sensed something odd as he read the CDC report. All five patients being homosexual could not be a coincidence. Absent a logical explanation, Fauci conjectured the men had ingested a drug that weakened their immune systems. Still, he sensed something amiss and lodged the report's findings in the back of his mind. So when another CDC dispatch came across his desk a month later, stating that twenty-six men—all gay—had been treated in three different cities for that same rare strain of pneumonia, Fauci rightly assumed the start of an epidemic. Captivated by the prospect of studying a brand-new pathogen, he dropped his research into autoimmune disorders to pursue studying the virus full-time.

In his memoir, Fauci does not address the origins of HIV directly, but does attribute its spread to increased acceptance of gay sex in the 1970s.[5] That explanation is likely insufficient as homosexuality had been present for millennia, whether in the open or behind closed doors. Theories regarding the origins of HIV range from it being a deliberately engineered bioweapon to the product of a romance between man and ape. The most widely accepted theory is that hunters handling infected meat in Central Africa transferred the virus from animals to humans. Whatever its roots, the structure and characteristics of HIV primed it to wreak havoc on post-sexual revolution America.[6]

By the early 1980s, the AIDS epidemic (or "pandemic," depending on how you want to classify the scope of the disease at the time) was in full swing. Having begun in gay communities, the virus had jumped to women and straight men through heterosexual sex and needle sharing. Thousands of grievously ill patients with wrecked immune systems often suffering from multiple different diseases at once, flooded hospitals around the country.

Fauci treated thousands of them at the NIH, describing the period as the "darkest years" of his medical career. With AIDS being a newly discovered disease, there were few available treatments, let alone an understanding of its root cause. Opportunistic infections, meaning additional pathogens patients contracted as a result of their collapsed immune systems, like pneumonia or tuberculosis, could be treated on the spot—but doing so was like playing a game of whack-a-mole. As soon as one illness was treated, another would pop up elsewhere. Fauci watched an AIDS patient progressively go blind as a result of contracting cytomegalovirus—normally a mild flu-like condition, rendered devastating by HIV.

The immense amount of stress Fauci endured during those first hellish days of the epidemic, comprised the most formative of his life. In working eighteen-hour days, month after month, Fauci watched as thousands of mostly young men helplessly succumbed to the mysterious disease. The stress and long hours even contributed to the end of his first marriage. Watching the effects of illness at that intimate level, instilled a life-long belief that anything was justified to end this kind of suffering—*anything.*[7]

Finally, in 1984, French scientists discovered the HIV virus, beginning a path forward out of the darkness. Drugs and treatments could be developed targeting the cause of AIDS itself. That same year, Fauci was appointed head of the NIAID, becoming the youngest ever leader of an NIH department—a position he would occupy until 2022. He now possessed authority to request more funding and direct those resources as he saw fit. Given Fauci's background and passion (bordering on obsession) to fight AIDS, he pivoted a huge portion of the department's budget towards battling the disease.[8]

Dozens of drugs began making their way through the federal government's long and torturous approval pipeline. The first to show promise was a drug formerly used to treat cancer called azidothymidine, or AZT. In clinical trials, groups administered AZT performed significantly better than placebo groups. Its promise was short-lived, however. In replicating billions of times a day, viruses inevitably make mistakes, permeating mutations. Some of these mutations confer increased immunity toward given treatments—which is what occurred with AZT. Initially successful, HIV mutated to obviate the therapeutic effects of the drug. In addition, much like how chemotherapy can kill cancer patients, AZT occasionally poisoned its recipients. Of course, scientists like Fauci were only figuring this out in the moment, engaged in a desperate fight against a plague that was infecting hundreds of thousands in the United States alone.

After years of trial and error, researchers discovered that a combination of drugs worked best. By the mid-1990s, administering three antiretroviral drugs—indinavir, AZT, and 3TC—dramatically reduced the virus's ability to replicate in the human body. The drug combination stymied replication to such an extent that virion levels became undetectable, rendering it incapable of transmitting to other carriers. The joint efforts of clinicians at institutes like NIAID and ingenuity of pharmaceutical companies to render AIDS a relatively benign disease is truly one of the greatest achievements in medical history.[9]

However, like his response to COVID, Fauci's treatment of AIDS did not go without criticism. In a 1988 open letter to the *San Francisco Examiner*, gay rights activists called Fauci a "murderer" and advocated "he be put before a firing squad." These overzealous critics argued Fauci had sentenced thousands of AIDS patients to their death by treating them with ineffective and occasionally toxic drugs like AZT. Additionally, the standard scientific practice of administering a control group with placebos was lambasted as murder.[10]

Yes, conducting trials of drugs without proved efficacy is a regrettable part of drug research. It is also true that giving control groups a placebo when they could potentially benefit from the real thing is a regrettable part of scientific research. In essence, the first group of patients to be treated for HIV were sacrificed so that contracting the virus today is not a death sentence given the plethora of effective treatments. The first wave of infantry at Normandy was dramatically more likely to be killed than the twenty-seventh wave. Our world is built on sacrifice.

Testing drugs like AZT on control groups was an unfortunate necessity of scientific progress.[11]

Fauci understood that despite criticism from radicalized activists, his laboratory was doing the right thing in developing long-term, effective treatments by following the scientific method. Furthermore, by abandoning his secure field studying rare autoimmune disorders to tackle the biggest pandemic of the era, Fauci chose to enter the highest "arena" of medical research. In the arena, whether in medicine or in life, its members open themselves to criticism. Fauci conducted legitimate scientific research to produce effective drugs, at scale, to save lives. For this, he should not be condemned.

Given HIV imbeds itself into an individual's DNA upon exposure, vaccine development was impossible. Therefore, developing therapeutic drugs to boost T-cell levels and diminish the effect of virions was the only path forward. A process which required extensive experimentation and unfortunately—sacrifice. Thankfully, Fauci did not go into the Castro District of San Francisco, lockdown the population and forbid interactions between residents to combat the disease. Though, given his actions forty years later, he may have thought about it.

Understanding Fauci's experience combating the AIDS pandemic is the first step to understanding his inane reaction to COVID-19. In treating thousands of AIDS patients, Fauci experienced a phenomenon common to many medical practitioners. Dramatically improving or even saving the lives of patients almost always, and understandably, ends with heart-felt expressions of gratitude and even love for what the doctor has accomplished. In restoring life to others, doctors take on the ultimate role of savior. In many cases, their expertise and ability, earned through years upon years of schooling, renders them the only ones capable of occupying this role. Most doctors deservedly view themselves as performing a tremendous social good for humanity and take pride in their work. However, for some, they develop an arrogance, believing they are above criticism and their actions beyond reproach—otherwise known as the "God complex."[12]

Not reserved to physicians, but infamously present in their field, a God complex can sanction any kind of behavior, however misguided or inconsiderate. *"I am the only one capable of saving this person's life. Your opinion doesn't matter. I should be given full reign to do whatever I want in order to save this person's life."* Living under this delusion, doctors are known to become unhinged, abandoning rationality and even hurting their patients as a result. The tradeoff is that afflicted physicians often feel immense pain and guilt when their patients do worsen or die. Being God is not all it's cracked up to be.[13]

Indeed, this author does not claim to fully understand the burden of physicians nor empathize completely with the immense responsibility they hold in saving the lives of others. That being said, doctoral expertise or position does not justify a full-fledged surrender to their recommendations, especially when they are exhibiting delusions of grandeur. Even more importantly, if their prescriptions influence fields outside of medicine—as occurred during the COVID-19 pandemic—then society holds absolutely no obligation to listen to them.

During the pandemic, Fauci and other medical professionals understood themselves as beyond reproach. Instead of engaging in rational discourse regarding their recommendations, supporting them with reason and logic, criticism was perceived as an afront to their endowed role as saviors.

While a psychological evaluation has not been conducted on the man, even a cursory examination of Fauci's actions during COVID suggest he suffered from a God complex. "Attacks on me, quite frankly, are attacks on science," the good doctor said. In responding to criticism from Congress that his recommendations to prohibit society from functioning were unwise, he countered that "science and truth are being attacked." It is not a stretch to claim these delusions germinated during his years treating AIDS patients. He emerged a medical hero having helped save the lives of thousands. Can he be blamed for believing his recommendations should be unquestionably followed?[14]

Another development during the AIDS pandemic which later hampered efforts to reign in Fauci was "HIV denialism." Spearheaded by molecular biologist and professor at the University of California at Berkeley Peter Duesberg, denialists argued HIV was benign and did not cause AIDS. Rather, particular lifestyles choices like drug use and unprotected sex weakened immune systems precipitating the syndrome. Additionally, Duesberg and others contended that a collection of viruses caused symptoms associated with AIDS, and that HIV was only one contributory cause to a broader problem. There is overwhelming scientific evidence to suggest this is incorrect.[15]

While skepticism toward "overwhelming scientific evidence" is absolutely necessary—as has been the case in regard to climate change—there are some facts beyond refutation. One is that there exists an enveloped, capsid-structured retrovirus that attacks immune system cells, plummeting T-cell levels within the human body. The virus is called HIV and can be physically observed under a microscope. Next, AIDS patients have consistently been observed to have contracted the HIV virus, and populations with high HIV prevalence exhibit high rates of AIDS-related illnesses. Finally, and most evidentiary of the link between the two, is that antiviral drugs designed to specifically target HIV virions diminish the symptoms of AIDS.[16]

Again, healthy skepticism toward scientific evidence is prudent. Lobbing the oft-cited *"according to this study I read,"* is woefully insufficient in supporting an argument. In 2016, researchers could not replicate the findings of seventy percent of "peer-reviewed studies." Replicability is a cornerstone of the scientific method, and if findings cannot be repeated then they should not be taken seriously—let alone influence public policy.[17]

For example, drawing a causal link between carbon emissions and the enormously complex processes that regulate planetary temperatures is dubious at best. Studies "confirming" the effects of climate change have consistently been unreplicable by other researchers. Nonetheless, alarmists relish citing *"overwhelming scientific evidence"* or the idea that *"97% of all researchers agree"* in refuting anyone who dares question the climate change narrative. In doing so, they trample over the tenets of scientific thought which prides skepticism and fallibility. A process which entreats its adherents to constantly question consensus and re-test hypotheses again and again. Unwavering agreement among researchers is almost always a tell-tale sign that their particular field is corrupted and untethered from true scientific inquiry.[18]

However, molecular science differs from climate science in that the former has progressed to where viruses can be consistently observed to target specific cells which spawn certain symptoms. Anyone with a microscope can observe HIV virions attacking immune system cells, triggering AIDS. Refuting this causal link contradicts basic facts about microbiology—much like how claiming masks and social distancing protocols mitigated the transmission of aerosolized coronaviruses.

Fortunately (for the rest of the world), only in South Africa did this fallacious conception about AIDS actually influence public policy. Duesberg managed to convince the South African president to ban the use of antiviral drugs in treating AIDS. Not surprisingly, from 2000 to 2005, this led to over 330,000 entirely avoidable deaths within the country. Nonetheless, Fauci was forced to debate skeptics like Duesberg in public which only elevated denialist theories.[19]

The experience would cast a long shadow over how Fauci would handle questioning towards his own recommendations during the COVID-19 pandemic. Fauci conflated the skepticism he received in recommending lockdowns with skepticism towards microbiological facts about COVID itself—as what had occurred with HIV denialism. Fauci believed that in refuting lockdown skeptics he was averting the kind of disaster he had witnessed in South Africa. In reality, the two cases were very different, requiring dramatically different approaches. Nevertheless, Fauci would carry these experiences with him throughout his career. A life studying, battling, and treating infectious diseases would come to its grand summation in 2020.[20]

By then Fauci was possibly the most famous doctor in the world and inarguably its most revered immunologist. While not eradicated, he had led efforts to suppress AIDS, the biggest pandemic in modern U.S. history. In the interim, he nearly accomplished the same in the developing world along with working to diminish the effects of influenza, Ebola, and SARS. He worked on biodefense efforts following 9/11, helped treat wounded soldiers in Iraq, and streamlined vaccine production globally. In total, he ascended to the peak of the United States public health establishment—frequently giving briefings to lawmakers on Capitol Hill and personally advising every president from Reagan to Obama. The diminutive Brooklyn boy raised by pharmacists had risen to the highest echelons of power.[21]

***

On January 20th, 2020, a thirty-five-year-old man returning home from Wuhan to Tacoma, Washington, tested positive for COVID-19—marking the first confirmed case of the disease in the United States. However, subsequent studies indicate the virus likely entered the country in December the previous year. Regardless, the aerosolized pathogen had made its inevitable arrival onto U.S. shores.[22]

As cases started to spread, the White House called the nation's foremost infectious disease expert for advice. In a subsequent meeting with top national security officials and representatives from Health and Human Services, Fauci deliberated how to respond. Halfway through the meeting, a fellow New Yorker walked in the door. "Anthony," President Trump said, "you are really a famous guy. My good friend Lou Dobbs told me that you are the one of the smartest, knowledgeable, and most outstanding persons he knows." Thus began one of the most influential partnerships in history.[23]

The Queens native, enamored by Fauci's glamorous reputation gleaned from having "defeated" AIDS, was keen to glom onto anything the Brooklynite recommended. Trump no doubt salivated at a chance to notch another "victory" to an overall successful presidency. *The stock market up 8000 points, ISIS defeated, and COVID eradicated!*

More importantly, the former reality TV show host is an avowed germaphobe and member of that rather sizable group that reacts viscerally at the thought of microscopic bugs infecting their body. Additionally, being seventy-four years old at the time, the specter of sickness is all the more pronounced. A fact which should not be underestimated in assessing why Trump, who campaigned on *not* kowtowing to Beltway "expert" recommendations, abandoned all semblance of democratic oversight and relinquished total authority to them.[24]

Laughably, his deranged critics, eternally compelled to criticize his every action, ended up criticizing actions *they advocated for.* By ceding decision-making powers to Fauci, Trump sanctioned lockdowns and "strongly reacted" to the virus. Another dose of irony which only functions to further elicit the absurdity of the era.[25]

Nonetheless, Fauci initially cautioned prudence, advising the president that with "identification, isolation, and contact tracing" the virus could be adequately contained. Furthermore, in a radio interview, Fauci told the public: "The American people should not be worried or frightened by this. It's a very, very low risk to the United States." Shortly thereafter, Dr. Nancy Messonnier, the director of the National Center for Immunization and Respiratory Diseases at the CDC, stated that "the current risk from this virus to the general public was low." If only this kind of thinking prevailed.[26]

Instead, the alarmist and fatalistic began to screech. Dr. Michael Osterholm, the same academic who complimented the draconian measures of the Chinese Communist Party in combating the virus, predicted millions would die in the United States. Senator Tom Cotton, in a tone befitting most senators keen to make news headlines, said that "this coronavirus is a catastrophe on the scale of Chernobyl." Nonetheless, most rational people, including a majority of immunologists and epidemiologists initially recommended discretion. They recognized the clear signs of another SARS-like coronavirus—a threat which mandated additional protections for the vulnerable but not the obliteration of modern society.[27]

Before we celebrate Fauci too fervently for his initial prudence, it must be acknowledged that he largely acted this way to avoid exposing his own role in causing the debacle. Starting in 2019, the NIAID funneled a total of $7.4 million dollars to facilitate research on bat coronaviruses at the Wuhan Institute of Virology. By that time, gain-of-function research—which literally means to add an ability to a virus—was a major industry within public health. Billions of dollars had flowed from federal coffers to facilitate its development over the preceding two decades.[28]

Fauci's own role in making viruses more transmissible to humans kicked off in 2000, when he approved funding for a study at the Utrecht University in the Netherlands entitled, "Retargeting of Coronavirus by Substitution of the Spike Glycoprotein Ectodomain: Crossing the Host Cell Species Barrier." An innocuous way of describing how a coronavirus was rendered capable of infecting other animals. This was only the beginning.[29]

The rationale for gain-of-function research is that in augmenting the properties of viruses, researchers gain an increased understanding of their transmission and lethality mechanism. Thus, if an outbreak ever does occur, society is better prepared to obviate their effects and develop effective treatments. For example, if through gain-of-function research, a virus is discovered to be only one or two mutations away from becoming airborne, then public health authorities can direct efforts to monitor its naturally occurring version. Additionally, within the laboratory setting, prospective vaccines can be administered to animal test subjects to determine efficacy. The obvious risk is that an augmented pathogen finds its way out of the laboratory and into the general public. If that occurs, taxpayers have now funded a life-threatening problem that was previously of no concern. Therefore, only the highest-grade security laboratories are authorized to conduct this kind of research.[30]

Some, most notably Robert F. Kennedy Jr., have suggested gain-of-function research is merely a cover-up for Fauci's deliberate creation of bioweapons in concert with the CIA. The theory, no doubt influenced by his uncle's suspected killing by the organization, posits that billions of dollars spent on studying infectious pathogens is actually a front for bioweapons development. Kennedy further suggests that these covert operations support population control programs.[31]

While he certainly has grounds to criticize gain-of-function research as a threat to humanity and is rightfully concerned about clandestine operations by the CIA, RFK Jr's claims connecting Fauci to bioweapon development are tenuous at best. Hanlon's Razor entreats us to not attribute malice to what is adequately explained by stupidity. Here, it is imminently practical.

The proliferation of gain-of-function research is rooted in arrogance within the public health establishment to believe the benefits accrued by making diseases *more* infectious outweigh the risks posed by their accidental release. To claim this research is conducted to build bioweapons to release on foreign and domestic populations falls closer to the realm of conspiracy than real fact. Hubris, not malevolence is to blame. Nonetheless, criticizing gain-of-function research as it relates to COVID-19 is highly justified.[32]

RFK Jr's work on the subject is phenomenal and readers are encouraged to read his book *The Wuhan Cover-Up: And the Terrifying Bioweapons Arms Race*. This author simply finds his conclusions that gain-of-function research feigned a CIA bioweapons development program as dubious. Kennedy's evidence primarily derives from a 1974 Henry Kissinger-authored study purporting to advocate for population control measures to avert a Malthusian calamity. Besides from a lack of source material, suggestions that the CIA aims to release biological diseases onto its population are circumstantial at best, and delusional at worst. Furthermore, if COVID-19 was somehow a deliberate attempt at population control, it did a lousy job. Global population numbers continued to rise during the pandemic and do so currently.[33]

Again, crafting a lethal pathogen to kill massive amounts of people is much harder than it may appear. The biological realities of a virus killing millions necessitate that its carriers not only live long enough to spread it, but somehow feel healthy enough to mingle in public with other people (which is difficult to imagine if the virus is seriously lethal). Therefore, a truly civilization-threatening pathogen would need to bypass our robust immune systems, yet also exhibit an extremely high incubation period, allowing carriers to spread the disease before an onset of symptoms.

Thus, this type of virus emerging in the twenty-first century would need to fulfill the following conditions simultaneously:

1) Spread via respiratory droplets and/or aerosolized particles like COVID or influenza. If it spreads via blood or semen like HIV, then it simply lacks the transmission mechanism necessary to transmit widely.

2) Possess an extremely high incubation period. In other words, pathogen vectors would need to be contagious *and* experiencing relatively benign symptoms so they feel comfortable interacting with others, *yet* die later on. Which brings us to:

3) Exhibit extremely high lethality. The absolute juggernaut of all respiratory viruses the likes of which the world has never seen, capable of not only compromising healthy immune systems (COVID could not) *and* be invulnerable towards the vast cornucopia of antivirals, modern sanitation practices and vaccines available to the modern world, would need to emerge.

The following are the probabilities for each of those respective conditions calculated by public health models:[34]

#1: **10%**
#2: **1%**
#3 **0.1%**

By multiplying each of the three individual probabilities together, we calculate the probability that *all three* are fulfilled simultaneously—which is: **0.000001%.**

This point does not need to be belabored, but it is obvious that of all the potential catastrophes facing our world, a deadly pandemic ranks near the bottom in terms of probability. Regardless, fear-mongering about its manifestation fueled obscene spending levels for infectious disease research, most notably on gain-of-function.

How else could governments justify funding high-risk experiments on viral genomes if not to avert some kind of apocalyptic calamity? Saying "infectious diseases are a concern but not that serious given the biological realities of transmission, incubation periods, and lethality" does not make for a riveting Ted Talk nor unleashes massive outflows of taxpayer money to your laboratory. Gain-of-function research would undoubtedly not have received nearly the level of funding it did, if researchers took that approach in grant requests.

Instead, the prospect of a deadly pandemic has been totally blown out of proportion. If we lived in the 14th century or even 1918, then extreme caution would be warranted given the desert of effective treatments and sanitation practices. Thanks to progressive reforms in the early twentieth century and advances in antiviral treatments, however, the prospect of a massively lethal pandemic is negligible. Yet, the media-hypochondriatic complex continues to project a fear-mongering brand of nonsense, scaring the public into thinking of pandemics as apocalyptic likelihoods.[35]

A belief which managed to infect the highest echelons of American politics during the mid-2000s. Amazingly, but not that surprisingly, part of the path to senselessly restricting the freedom of millions of Americans runs through the 43rd President. "All good things are connected to George W. Bush." Or so the saying goes, right?

In 2005, a White House staffer handed the former Texas governor a copy of John Barry's *The Great Influenza*—a history of the 1918 Spanish Flu pandemic. After reading the book, apparently not too closely considering Barry admits that nothing much can be done to combat the spread of infectious disease, the president became fixated on the threat.[36] Despite tremendous advances in sanitation and medicine since 1918, Bush deemed the prospect of forty to sixty million dying from respiratory illness a pressing issue. The president then did what everyone does after reading their first book on a subject and begged Congress for seven billion dollars.[37]

Of course, asking Congress for money is rarely denied (as after all, it is not their money) and funds were dutifully allocated for an anti-pandemic task force. Read a book, become fixated on an obscure contingency, ask for and receive seven billion dollars. Just how the Founders wrote it up! Here begins the country's slide away from sensible targeted responses against infectious illnesses to the indiscriminate restriction of entire populations with total disregard for their rights.

The task force's final report recommended innocuous and benign-sounding prescriptions like "where appropriate, use governmental authorities to limit non-essential movement of people, goods and services into and out of areas where an outbreak occurs." In other words, forcibly confine people inside their homes and restrict their freedoms. Another recommendation counseled, "social distancing measures, limitations on gatherings, or quarantine authority may be an appropriate public health intervention." Said differently, impose draconian lockdowns on citizens for as long as authorities deem necessary.[38]

The principal author of the report, Richard Hatchett, CEO of the Coalition for Epidemic Preparedness Innovations (CEPI), cited the 1918 Spanish Flu as evidence for implementing authoritarian public health measures. Hatchett argued that "social distancing, infection control, and travel restrictions" had been "relatively successful" in combating the early-twentieth century pandemic—despite admitting a "retrospective assessment is difficult."[39]

It is a good joke that the Spanish Flu would be cited as a threat to modern society. Anthony Fauci admitted as much in a 2008 study which found the majority of deaths during that pandemic resulted from bacterial infection, which is easily treated by modern-day antibiotics. Moreover, a 2000 study by Johns Hopkins and the CDC found that deaths from respiratory diseases had plummeted by 99.7% over the last century.[40]

Nonetheless, funding for anti-pandemic initiatives accelerated, spearheaded by gain-of-function research. NIAID alone received over $2.1 billion to genetically alter pathogens. Additionally, by 2005, Fauci had convinced President Bush to build four maximum-level biosecurity (BSL-4) labs and thirteen other BSL-3 labs across the country to facilitate gain-of-function studies.[41]

Even at the time, the practice was not without its critics. *The New York Times* editorial board—of all institutions—questioned the research in 2012, stating the "consequences, should the virus escape, are too devastating to risk" and recommended the programs be terminated. Given *The New York Times* editorial board did not possess any degrees in microbiology or virology, their recommendations were justly ignored.[42]

Researchers in Australia modified a strain of mousepox, enabling it to bypass natural immune system responses, dramatically increasing its lethality. Gary Nabel at the NIH, in not exactly a revelatory statement, labeled the news a "concern."[43]

In 2013, Lynn C. Klotz, a senior fellow at the Center for Arms Control and Non-Proliferation authored a "likelihood-weighted consequence analysis" which concluded gain-of-function research would precipitate thousands of deaths. Klotz further stated its risks did not outweigh its potential benefits. A prescient remark to say the least! While likelihood-weighted consequence analyses should not be necessary to prove genetically altering infectious diseases to increase their transmissibility and lethality is suboptimal to the human experience—we applaud the effort.[44]

Of course, others hailed and defended the practice—most notably Drs. Fauci and Francis Collins. In an opinion piece in *The Washington Post* published in 2011, the two co-wrote that "safeguarding against the potential accidental release or deliberate misuse of laboratory pathogens is imperative" and justified the risk based on the "important information and insights" gained from experimenting on "potentially dangerous virus[es] in the laboratory." Fauci further assured the public that only "very small amounts of [viral] material" would be housed inside BSL laboratories. Though, correct me if I'm wrong given my lack of a Ph.D. in Virology, but viruses *replicate*. Meaning, the escape of only one virion enables its potential proliferation throughout all of society. Fauci's comforting remark is akin to handing a gun with only one bullet loaded in its chamber to a toddler, justifying the act because *"the gun only has one bullet."*[45]

The information campaign to persuade policymakers continued unabated. In 2015, the National Institutes of Health published a report claiming that gain-of-function research "has made enormous contributions to the understanding of disease and the development of cures," but that its benefits "may be long-term and their value not immediately evident." The report is filled with "likelihoods" and "future possibilities" and all the hallmarks of an unproven and highly dubious, taxpayer-funded program.

In its chapter on "Biosafety and Biosecurity," the authors brashly claimed that "types of risks...had not been well defined" for gain-of-function research. Does the accidental release of viruses genetically modified to become more infectious to human beings require further definition? Its authors admit that existing "guidance, regulation and infrastructure" to address these risks was "not perfect." Is that good enough? The release of even one highly contagious virus would render the entire enterprise counterproductive.[46]

Thankfully, these infectious pathogens, rendered even more infectious through deliberate intervention, were handled safely and never escaped into the public. Human beings are known to never make mistakes and systems designed to safeguard against mishaps are foolproof and never compromised.

Just kidding! Pathogens escaped on numerous occasions. In 2014, three separate biosecurity breaches occurred across the country. At the Atlanta CDC, seventy-five staff members were exposed to mishandled live anthrax specimens. At that same laboratory, officials meaning to ship a relatively benign strain of avian influenza to an agricultural lab, instead sent samples of H5N1 avian influenza by mistake—a virus with a far higher lethality rate. The *oopsie* was only discovered when a flock of chickens at the USDA lab suddenly died. Less than a month later, Food and Drug Administration staff in Maryland, found smallpox cultures leaking out of an unsecured closet.[47]

Following these incidents, eleven members from the U.S. National Science Advisory Board for Biosecurity encouraged Dr. Fauci to improve BSL lab security measures. Demonstrating all the traits of an unelected bureaucrat, Fauci responded by firing all eleven members.[48] Nevertheless, media reports of the leaks along with lobbying from concerned researchers led President Obama to issue a moratorium on further gain-of-function research for "potential pandemic pathogens." A subsequently issued statement from the HHS advised that "the relative merits of gain-of-function experimental approaches must be compared ultimately to potentially safer approaches." The moratorium halted funding for viruses including MERS, SARS and other coronaviruses.[49] Thankfully, researchers abided by the presidential mandate, halting experimentation on lethal viruses until more robust safety procedures could be implemented. Unelected bureaucrats, funded by taxpayer money, followed the rule of law and never abused their position whatsoever. *Just kidding!* NIAID-funded researchers simply flouted the order and continued to conduct gain-of-function research on lethal viruses. Ralph S.

Baric at the University of North Carolina received tens of millions of dollars in NIAID grants to press on with improving the transmissibility of coronaviruses. Baric bemoaned that too few "studies to enhance transmissibility have been conducted using a coronavirus." To fill the void, Baric genetically modified bat and camel coronaviruses, enabling them to jump to intermediate hosts like rats and ferrets. One of his studies, conducted in 2015, examined the properties of SCH014, a viral strain collected by Wuhan Institute of Virology chief researcher Shi Zhengli from bats inside a cave in Yunnan province—a virus genetically related to COVID-19.[50]

Baric defended his research despite the moratorium arguing that Obama's mandate only barred new research from taking place. Since his research was "under way" it did not fall under the mandate. However, Obama's mandate did apply to on-going studies and Baric simply ignored the order.[51]

Opposition towards the moratorium from Dr. Fauci and others within the public health establishment eventually pressured the White House to remove the ban in 2016. Evaluating the safety of gain-of-function research would be left to "independent" reviews within the NIH, meaning Fauci himself was tasked with determining whether research should proceed. Not surprisingly, Baric and other gain-of-function researchers started receiving massive grants once again. By 2023, Baric alone had received NIH/NIAID grants totaling $215 million dollars.[52]

One such lab to receive funding was, of course, the Wuhan Institute of Virology. The massive, 32,000 sq ft, BSL-4 accredited laboratory took eleven years to build, finally finishing in 2015 at a cost of over $43 million. The NIH proceeded to fund at least sixty projects at the institute. As funding for gain-of-function became more controversial, the NIH funneled its support through an organization known as EcoHealth Alliance. Originally an environmental conservation group started in the 1970s, its leadership realized grant money was more liberally doled out for the study of infectious pathogens.[53]

Led by parasitologist Peter Daszak, the group began collaborating with Shi Zhengli at WIV in the early 2000s. Over the next fifteen years,

Daszak and Shi conducted numerous studies to enhance the transmissibility and lethality of infectious pathogens. The two jointly published a study entitled, "Bats are Natural Reservoirs of SARS-like Coronaviruses" in 2015. Daszak even boasted to an interviewer that his team had discovered over one hundred novel coronaviruses in bat caves throughout Southern China.[54]

From 2018 to 2019, their crusade to render pathogens more infectious to human beings accelerated. Aided by NIH funds, Shi and Daszak increased the pathogenicity of coronaviruses by altering their spike proteins to more readily bind to human receptors. In January 2018, Shi published a study entitled "Evolutionary Mechanism of Bat SARS-like Coronavirus Adapted to Host Receptor Molecules and the Risk of Cross-Species Infection." In other words, using humanized mice, her team had successfully demonstrated the ability of SARS-like coronaviruses to bind to human ACE2 receptors. Take a wild guess at what receptor COVID-19 binds to.[55]

At the same time, safety protocols at the Wuhan lab were in dire straits. In early 2018, the US embassy in Beijing sent an unclassified cable to the State Department, warning of a "shortage of highly trained technicians and investigators required to safely operate a BSL-4 laboratory" within China. Later, when State Department officials arrived in Wuhan to inspect the lab, they were shocked at what they discovered. The central air conditioning system was in disrepair, allowing poor circulation to leave viral particles suspended in the air. More importantly, they learned that Shi Zhengli had discovered SARS-like coronaviruses in bats could bind to ACE2 receptors in humans.[56]

What happened next should not require repetition. One of the coronaviruses Shi and Daszak experimented on escaped from the lab. End of story. It happened before with other viruses, as has been described, and it happened again. The Wuhan Institute of Virology in conjunction with the National Institutes of Health and EcoHealth Alliance accumulated samples of bat coronaviruses from nearby caves, returned them to the lab and genetically modified them to be more transmissible to humans. Whether by virtue of inadequate air ventilation or via other means, one of those modified viruses infected a lab worker, who later transmitted the disease to residents in Wuhan—sparking an epidemic.

Once the pandemic was in full swing, its perpetrators initiated the cover-up. Gain-of-function zealots like Daszak realized investigators would soon connect the dots back to their research in Wuhan, requiring a preemptive information campaign to deflect blame. In 2021, *The Intercept* described Peter Daszak as a "key voice in the search for COVID-19's origins." This is like when O.J. Simpson began a manhunt to find the killer of Nicole Simpson. Nevertheless, Daszak dutifully labeled any suggestion that the virus may have leaked from WIV a "conspiracy theory" and mobilized a legion of compliant scientists to support his claim.[57]

Their behavior is understandable as admitting the virus leaked from WIV would have justified nearly two decades of opposition towards their research. Billions of dollars in American taxpayer money could not be concluded to have sparked the very disaster currently underway. Thus, Daszak and others wisely perceived that labeling suggestions COVID leaked from WIV a "conspiracy theory" and "racist" capitalized on Western anxieties about misinformation and xenophobia.

On February 3, 2020, Shi published a study in *Nature* stating that COVID was 79.6% genetically identical to SARS. More importantly, in an effort to somehow refute allegations that COVID did not originate from her laboratory, she proclaimed SARS-CoV-2 was *only* 96.2% identical to a bat coronavirus her team had collected from a Mojiang cave in Yunnan province. Imagine that! It is almost as if they were experimenting on that coronavirus to make it more transmissible to humans at the same lab that COVID escaped from![58]

A few days later on WeChat, Shi posted: "COVID-19 is nature's punishment for uncivilized living habits of human beings. I, Shi Zhengli, use my life to guarantee that it has nothing to do with our lab." If by "uncivilized living habits of human beings," she means her lab's inability to contain the virus, *then she was spot on!* Apparently that was enough for *The New York Times* as they labeled the claim it leaked from her lab a "fringe theory" following her statement. In 2024, to finally squash those dangerous rumors, Shi provided "samples" from her lab which were genetically unrelated to COVID. *Case closed!*[59]

***

Unfortunately, part of the reason why the lab-leak hypothesis was deemed a "conspiracy theory" derives from its utter incomprehensibility. Indeed, it sounds like a wacky, tin-foil hat idea to suggest viruses were deliberately made more transmissible to humans—like something the Umbrella Corporation from *Resident Evil* would do.

Nonetheless, it happened and there are legitimate, though highly tenuous reasons for why the practice began. Indeed, just as concerned members of the scientific community warned of its risks outweighing its benefits, the COVID-19 pandemic proved the unacceptability of conducting gain-of-function research. It is an understatement to conclude the pandemic's immense economic and social strains outweighed any benefits accrued by storing and genetically modifying infectious diseases.

It is possible that Fauci, Daszak, Shi, and other gain-of-function zealots never imagined how insanely world governments would react to the leak of one of their viruses. It is possible they assumed its "risk" was that society would react similarly to the SARS pandemic, isolating vulnerable populations until herd immunity was developed—with life continuing relatively unaffected. Yet, the unique blend of COVID emerging in an authoritarian country combined with RT-PCR and rapid digital communication-induced panic mutated the threat beyond all comprehension. In their defense, the subsequent hysteria was stoked by media professionals and hypochondriacs escalating the virus's escape into a dystopian hellscape. As described above, Fauci even attempted to quell fears about the virus during the outbreak's early days.

Though once it became clear that COVID had morphed into a global catastrophe with investigators inevitably tracing its origin back to gain-of-function research in Wuhan, Fauci quickly changed his tune. Ultimately, their arrogance in accelerating gain-of-function research despite protestations by the medical community is disdainful and responsible for the origins of the pandemic.

In May 2021, Fauci testified before Congress that "the NIH has not ever and does not now fund gain-of-function research at the Wuhan Institute of Virology." A clear case of felony perjury, carrying up to five years in prison. Of course, perjury is hardly ever prosecuted and given Fauci's elite position within the country, he was never going to be charged for this. President Biden's outgoing pardon of Fauci ensured he never will.[60]

SUMMER

"Look at what I did to this city with a few drums of gasoline and a couple of bullets."

– Joker, The Dark Knight[1]

Here is not the place to address the multi-faceted and complex dynamic that is race in America. This is a history of the COVID-19 pandemic. Yet, oddly enough, race eternally intertwined itself with the respiratory disease during the summer of 2020. In not exactly a shocking development, confining millions inside their homes, barring their employment and prohibiting their recreation precipitated mass rioting.

Nevertheless, in a verbal jiu-jitsu move rivaling that of denying COVID's leak from the Wuhan Institute of Virology, some have attempted to separate the two. For example, the "Black Lives Matter protests" Wikipedia page entry contains only *two* references to COVID-19, and in no way links riots to lockdowns. Existing histories of that summer focus exclusively on the role of police brutality in America and past injustices inflicted upon blacks. Protests were deemed a natural outcrop of centuries of bottled-up rage and frustration against systemic racism—with the death or murder of George Floyd (depending on how you interpret toxicology reports) credited with igniting the fuse. The pandemic receives a tangential mention at most, resigned to an interesting, but ultimately unimportant coincidence. However, doing so is an insult to common sense and gross misuse of the word.[2]

Again, pandemics usually do not concern differing skin colors. Of the increasingly few color-blind characteristics of our race-obsessed society—viruses are one of them. They do not discriminate or segregate. They merely replicate. Yet, any history of the pandemic is incomplete without the inclusion of BLM riots in its narrative. The looting, burning and effacement of nearly every one of America's major cities will always be a direct consequence of restricting the freedom of millions over a cold virus.

Where do we begin? The lynching of innocent blacks in the Deep South? A shameful compromise between southern and northern delegates at the Constitutional Convention? The offering of a counterfeit twenty-dollar bill in Minneapolis? There is no shortage of starting points for the enormously complex and dark history of race relations in the United States. Here, it is enough to say that this author's generation grew up believing a "post-racial" future was upon us. Framed photos of Reverend Martin Luther King Jr. in classrooms, school presentations extolling the accomplishments of famous African Americans during Black History Month, and endless reminders of the sins of White America, endowed this author's generation with the idea that racism was behind us. Students were taught and believed in judging others not by the "color of their skin but by the content of their character."[3]

Of course, whites would still need to feel guilty. Screenings of films like *Amistad, Selma,* and *12 Years a Slave,* evoked the ritual cringing of white people and a reminder of what *they* did. Whenever it came time for classrooms to read that heinous and despicable disyllabic word in *To Kill a Mockingbird* on page 174, teachers would take over reading aloud and dutifully pause at its evocation.[4] The class would remain paralyzed in ponderous silence gazing at the most infamous word in the English language. Reading commenced but with the significance of its omission imbedded into the middle schooler's mind—left with the indelible impression that it would remain off-limits for particular skin colors.

In total, these exercises in self-flagellation affected to instill guilt and remorse within whites. And rightfully so! The whips and chains, the auction blocks, the angry mobs with torches, and the pain of over two hundred years of subjugation were constant reminders of America's original sin. Not reckoning with that past is an insult to history's greatest utility: learning what *not* to do! Weaving race with legal status was a horrible and asinine relic of the pre-modern era, never to be repeated again.

Nonetheless, it was *over*. The nation had waged a bloody civil war to liberate its slaves. Racists and demagogues continued to discriminate for another hundred years but the heroics of civil rights leaders ensured their defeat. The partnership between Lyndon Johnson and MLK to pass the

Civil Rights Act of 1964 symbolizing the reconciliation between the two races. The struggle culminated in 2008 with the election of the nation's first black president, Barack Obama. At last, America had arrived. Blacks ruled the sports world, the music industry, and now the White House. There were no longer any legal, cultural or political barriers to their ascension.[5]

Yes, a southside Chicago gas station clerk may double-take at their entry into his store. Yes, that homeowner in the Pacific Palisades sporting an inverted bob may stare a bit too long at certain members of the population driving through. But similar experiences would plague other races too. Whites would not feel all that welcomed in a fraternity at Morehouse College nor be embraced at a Black Panthers meeting. It is perfectly natural that certain spaces determine acceptance based on common understanding and experience—which is heavily dependent on race. The world is not a pharmaceutical drug commercial with members of disparate skin colors seamlessly mingling together *all the time.* However, it is absolutely unacceptable that entry into business, politics, the military or other fields devoted to societal advancement and public service be conditioned on race.

Whites legally disallowed blacks from entering certain hotels, shops, schools and other venues based solely on the color of their skin for centuries. This was wrong and rightfully abolished with the passage of the Civil Rights Act. Thankfully, the self-rejuvenating qualities of the American experiment had righted the wrongs of 1789 and realized the dream that "we are all created equal."

At last, American society seemed ready to judge others based solely on merit and ability. Yet, something curious happened. The pendulum had swung back to neutral but that was not enough. It was time for revenge! The Sauron eye of discrimination foisted on blacks for hundreds of years shifted its gaze to their historical oppressors. It was the White Man's turn to feel its fire. Affirmative action programs sprang up in universities across the country. Whites, but more so Asians, would confront tougher admission requirements based on their skin color. Of course, it was politically infeasible to call these programs "race-based admissions," but the effect was all the same.

Administrations justified the practice by claiming their campuses should reflect a broad range of backgrounds, cultures, and ways of life. They argued visitors should not confuse Harvard Square or the New Haven Green with Peking or Fudan University—which of course they would if admissions were based solely on test scores.[6]

Next, came "diversity quotas" in the business world, meaning black applicants were prioritized over their counterparts in hiring and promotion. "Diversity" simply becoming a codeword for "anti-white" and "anti-Asian." Furthermore, the Armed Forces, the organization tasked with winning the nation's wars, would prioritize persons of color for admittance into the officer corps. All of it justified under the premise of "leveling the playing field" and righting historical wrongs.[7]

Yet the playing field was already leveled! It had been illegal to discriminate based on race since 1964. Nonetheless, the nation that seemed so ready to ignore race had embraced its seduction once again. Simply switch the "preferred demographics" within any of these programs from black to white and one can only imagine what *The New York Times* or *Washington Post* would print incessantly about until they ended, and its perpetrators imprisoned.

The great tragedy of this regression to antebellum notions of race would be that the very individuals these programs claimed to help the most—competent persons of color—would be judged all the same. As an example, after graduating top of his class from Yale Law School, future Supreme Court Justice Clarence Thomas faced the unintended sting of affirmative action. The brilliant and highly capable Thomas struggled to find employment at the nation's top law firms. Employers assumed his admittance into Yale was predicated on race and so rejected his application. Furthermore, after finally being hired, Thomas reported his peers questioned his presence, believing skin color was primarily responsible for his employment.[8]

This is the colossal failure of it all. As long as affirmative action programs exist, even when a person of color obtains a high-ranking position based on innate merit and ability, others will assume race was responsible.[9]

Anyone who has even lightly engaged with the human race will know that there are idiots who happen to be white and there are idiots who happen to be black. There are imminently capable people who happen to be white and imminently capable people who happen to be black. Thus, as admissions criteria to college or professional work becomes further removed from merit and ability, the more it disadvantages everyone—regardless of skin color. Whether it be in eroding the quality of the organization or in fostering skepticism toward legitimately competent persons of color, the effect of affirmative action is corrosive.

Ironically, many twenty-first-century whites proudly claim that had they been alive during Jim Crow or the antebellum period, they would have "stood up against racism!" and lent aid to subjugated blacks.[10] If past performance is any predictor of future performance—but also retroactive, fantastical, historical performance as well—then whites who support affirmative action today would have been the ones cracking the whip and leading the lynch mob of yesterday.

Some argue affirmative action and other race-based admission programs simply rectify the historical record. Is it not justified that whites feel the same sting of racism they inflicted on others for so long? Is it not reasonable that we "get even?" History is replete with this moral dilemma. Should the Armenians start a genocide against the Turks? The Romans enslaved the Gauls, so should the French now enslave the Italians? Should the Jews repeat the Holocaust on modern-day Germans?

An influential carpenter from Nazareth reportedly said, "if anyone slaps you on the right cheek, turn to them the other cheek also."[11] But is it that simple? When someone is proven to have committed murder, most societies justly kill them in turn. Attacking someone is wrong but the victim's defense often requires a counterattack. There are no easy answers, and this book is not a treatise on ethics. It only chronicles what happened and leaves judgment to others. Here it is only sufficient to say that in America, the heinous and despicable were not abolished, but simply shifted onto another group.

The table-turning of racial discrimination would extend culturally. Starting in the 2010s, racial ideology ascended to dominate American mass media and popular culture.

In essence, experiencing art and entertainment became an exercise in white self-flagellation. The inevitable stooge or morally compromised individual in any television commercial was a white person. Whites overwhelmingly portrayed either imbeciles or irredeemable villains in film and TV. Pulitzer Prize finalist and feminist icon Joyce Carol Oates tweeted in 2022 that "the straight white male is the only category remaining for villains and awful people in fiction and film and popular culture." Furthermore, dramatized period pieces cast blacks to play historical whites like Thomas Jefferson or Queen Charolette; but the reverse was labeled "whitewashing." One can only imagine what *Slate* or *MSNBC* would crow about if Brad Pitt portrayed MLK—or Matthew McConaughey played Malcom X.[12]

Moreover, a subset of black cultural expression in music from the likes of Lil' Wayne, Lil' Uzi Vert, NBA Youngboy, and other modern-day Mozarts came to dominate the musical mainstream. In effect, white audiences were subjected to hearing constant utterances of the N-word, while remaining under restriction of using the word themselves.[13]

Which brings us to the most powerful word in the English language. Yes, how we use language shapes power relations within society. Those who dismiss its effects are blind to its influence. Maoist China banned the use of the words "individualism" and "democracy." Nazi Germany banned the Yiddish inflection of certain words. North Korea has banned the words "dictator," "human rights" or words associated with Western culture like "McDonald's" and "Coca-Cola." Linguistics professor Noam Chomsky wrote that "Language is not just words. It's a culture, a tradition, a unification of community, a whole history that creates what a community is. It's all embodied in a language."[14]

While the First Amendment negated the outright banning of whites using the N-word in the United States (which would also violate the Civil Rights Act), censors enforced the policy through other means. While not legally sanctioned—social ostracization, reputational and professional assassination and even *physical violence* were sanctioned to enforce the edict.[15]

Obviously, calling a person of color the N-word out of hatred or in an attempt to wound them is asinine and abhorrent. However, the social policy extended the ban toward whites using the word out of any conceivable discriminatory context or even *quoting someone else's* usage of the word. The custom to not use the word out of respect and acknowledgement for the horrible sins of slavery mutated into a mechanism for power and control completely untethered from its previous social utility.

As is dutifully followed in this book, whites are restricted to only referring to the slur as "the N-word" (despite the provocative tendencies of your author, even I dare not fully test the taboo). Other racial epithets like "cracker," "beaner" and "chink" could freely be said without anything resembling a similar backlash. There is no "C-word" or "B-word" for good reason.

Some extremist censors even argue using the term "black" or "blacks" by whites is unacceptable. Thus, the term "African American" became yet another tool of language control, even though calling U.S. citizens with ethnic roots to Italy or Ireland "Italian Americans" or "Irish Americans" would be ridiculous.[16]

While this book employs the language control mechanism of the "N-word" out of respect for its abhorrent use in the past, refusing to use the word "black" is an insult to the descriptive utility of language itself. Using the term is no more racist than calling those with ethnic roots to Europe as "whites" or those from Asia as "Asians" or those who work with wood as "carpenters." The further language diverges from its primary purpose of accurately describing reality, the more it corrupts.

The strict definition of "racism" is discrimination based on skin color. Therefore, a useful test to identify a racist is through the following thought experiment:

Imagine there is a white person floating through space in a pressurized and hermetically sealed box. The individual is freely drifting within the box, suspended in the middle by the environment's reduced gravity. The person's face is completely neutral, eyes open but mouth closed—their arms and legs dangle loosely. There are no thoughts, no feelings, and no signs of life except the regular autonomic systems of the heart and lungs operating normally. The person is alive but more closely resembles an inert object.

Now imagine this: a radio wave is projected into the individual's frontal lobe to initiate speech. Their diaphragm contracts and air is propelled into the trachea towards their vocal cords. As air passes through the throat, the radio wave provides instructions for how the vocal cords should manipulate pitch and tone. Further instructions are provided for how the raw sound should be conformed, dilated, and shaped by the person's tongue, lips, teeth, soft palate and jaw. The soft palate is engaged to create an "kn" sound. The lips wrap in a circle to push out the syllable "gur." The final result: a white person has said the N-word.

If someone considers this an act of racism; but not if the floating person's skin color is changed to black—then they are racist! This is not a moral judgment but merely a logical extension of the definition itself. No one would ever call the floating white person racist, you say?

Actress Gwyneth Paltrow was forced to apologize for tweeting the title of a Kanye West song, "N***** in Paris." The same occurred when a video of socialite Paris Hilton surfaced of her *quoting* the lyrics to a song which contained the N-word. In 2021, a white Rutgers Law School student was admonished by a *Washington Post* columnist for *quoting* the word in a legal opinion. The columnist concluded with the enlightened judgement of "Should you say the N-word? No, especially if you're not Black." Speech censors caught Papa John's founder John Schnatter *quoting* how KFC's Colonel Sanders used to call black people the N-word on a company conference call. As a result, he was forced to resign the chairmanship for a company he started. Does the prospect of extremists screeching about racism in response to the thought experiment sound all that implausible now?[17]

Observation is not judgment. Empirical claims merely describe the world. Normative ones describe how it should be. The preceding paragraphs are merely a set of empirical claims obvious to anyone who observes their surroundings. As an example, yours truly can remember discussing the intricacies of the United States federal budget with a fellow college undergrad. Possessing all the hallmarks of illumination and erudition a discussion between two twenty-year-olds is commonly known for, my peer told me that the American federal government spent the majority of its budget on military expenditure. All the American political system cared about was waging imperialistic war, she claimed. I uttered the demonstrable fact that nearly seventy percent of the U.S. annual budget, at the time, was devoted to "income transfers," meaning entitlements and handouts. She gasped and admonished me for the remark. I had merely stated a fact to her without judgment. Maybe I loved that fact! Maybe I thought it should be a hundred percent!

As an aside, the incessantly proclaimed statement that the United States is a capitalist dog-eat-dog world with no socialized institutions or income redistribution is laughable based on even a glance at federal and state expenditures. Indeed, Lenin and Marx would smile at a pie graph of the U.S. federal budget. Nonetheless, my peer had confused an empirical claim for a normative one. Similarly, the statements in this book, and especially in this chapter, are products of simple observation.[18]

How about another observation? We live in a race-based society—a race-based hierarchy. Anyone honest who has lived in and experienced societal relations in the United States will admit this. Saying so merely echoes what leftists and neo-Marxists have screamed about for decades. University of Maryland sociology professor and Black Lives Matter supporter Patricia Hill Collins wrote that, "what we have in the United States is a racial hierarchy."[19]

She is correct. Certain racial groups predominate certain employment and economic strata. The racial makeup of a construction site looks very different than the offices of Goldman Sachs. Driving through Bel Air then Compton; or visiting Delaware Country followed by eastside Philadelphia further evidences the point. Is each some kind of melting-pot, kumbaya, pharmaceutical commercial-looking utopia where races freely mix and picnic together? Obviously not, and when conservatives say that race does not matter in America, they are either willfully ignorant or deliberately lying.[20]

Race dominates this country and apparently always will. The move toward a post-racial future failed miserably. No sooner had the ink dried on the Civil Rights Act than affirmative action programs and diversity initiatives began to proliferate *en masse*. Indeed, it appears Americans are intent on dominating others based on race.

But point to a nation where this is not true. Point to a nation where different races effortlessly mingle together. Certainly not in Europe. The continent is currently engaged in a practical race war regarding the importation of millions of non-Europeans. In Eastern Europe, people separated by only a few hundred miles view their neighbors as sub-human and unworthy of even the smallest modicum of respect. Certainly not Asia. The Chinese and Japanese notoriously despise each other and both look down upon Southeast Asians. The Middle East? Give me a break. Is this right? Is it just? It's irrelevant. It is the truth of how human populations have behaved for millennia.[21]

In fact, the United States up until the 2010s was the closest to any political body in history to achieve something resembling racial harmony. A diverse coalition of voters, including white suburbanites, lifted Barack Obama to electoral victory in both 2008 and 2012. There are famous and successful U.S. black celebrities, athletes, and business leaders. Do the Chinese have any black movie stars? Do the Ecuadorians? Have the French ever elected a nonwhite president? How about the Russians or the Swedes or the Dutch? Europeans are fond of calling the United States a "racist country," but their own power structures are dominated by ethnic Europeans. How about Asia? Could you imagine a black woman or a Latino sitting in the National People's Congress of China or the National Diet of Japan?[22]

Thus, the United States is more accepting to different races than any other nation in history. Not because the country is some kind of idyllic color-blind paradise (it is far from perfect), but because the bar is set so low anywhere else in the world.

***

Having acknowledged this, we return to our narrative. The year is 1991 and Latasha Harlins, a fifteen-year-old black girl enters a liquor store in South Los Angeles. Harlins approaches the store counter to purchase a bottle of orange juice. Eyewitnesses testified that Harlins placed the orange juice in her knapsack and approached Korean shop owner Soon Ja Du with cash in hand.

Security tape footage is too grainy to definitively confirm that account but what followed is unmistakable. Du accuses Harlins of attempting to steal the orange juice and snatches her knapsack. After a brief physical altercation over the bag, Harlins gives up and leaves the bottle on the counter. As Harlins turns to exit the store, Du pulls out a .38-caliber revolver and shoots Harlins in the back of her head. A clear case of second-degree murder, typically punishable by between twenty years to life in prison.[23]

Nevertheless, a jury found Du guilty of voluntary manslaughter, an offense which carries a maximum sentence of only sixteen years in prison. In a final insult to justice and human dignity, Los Angeles County Superior Court judge Joyce Karlin deemed Du "not a threat to the community" and that "she knew a criminal when she sees one." Under that logic, she dropped the prison term and sentenced Du to five years' probation and 400 hours of community service. Du had gotten away with murder practically scot-free.[24]

No wonder race riots erupted in the city the following year. No wonder blacks felt the system was rigged against them. No wonder a majority-black L.A. jury acquitted O.J. Simpson of murder only four years later—as whites gasped in dismay baffled by its verdict.[25]

Meanwhile, thirteen-hundred miles away in Houston, a tall and gregarious young black man prepared to graduate high school. The 6'6" power forward and tight end once told a friend, "I want to touch the world." His name was George Perry Floyd Jr., and he would do just that.

After attending Texas A&M Kingsville on a basketball scholarship, Floyd dropped out and returned to the Third Ward neighborhood of southside Houston. He began a rap career under the pseudonym "Big Floyd," joining the hip-hop group Screwed Up Click. *The New York Times* described his rhymes as "purposeful," rapping about "choppin' blades—driving cars with oversize rims—and his Third Ward pride." However, the years would not be good to Big Floyd as he served eight jail terms on various charges including theft, drug possession with intent to sell, and aggravated robbery with a deadly weapon. In 2004, the Houston "justice" system sentenced Floyd to ten months in prison for a $10 drug transaction.[26]

After his release on separate charges for armed robbery ten years later, Floyd moved to Minneapolis for a fresh start. He became involved with the community, volunteering at his local church and working security for a Salvation Army homeless center. "He would regularly walk a couple of female co-workers out at night and make sure they got to their cars safely and securely. Just a big strong guy, but with a very tender side," a fellow co-worker said. Floyd later became a bouncer at a Latin dance club in February 2020 and had aspirations to obtain his commercial truck driving license.[27]

That same month, a mysterious respiratory virus began circulating through the American population. Instead of adopting a wait-and-see approach to determine its severity, Governor Tim Walz ensured Minnesotans would endure some of the most restrictive lockdown provisions in the entire country. Walz ordered all restaurants, gyms and other "non-essential" businesses shut down, limited indoor gatherings to no more than ten people and closed schools across the state. The future vice-presidential candidate even established a hotline enabling residents to report fellow citizens for not following social distance mandates.[28]

Walz's dictates closed Floyd's night club, forcing his boss to lay him off. Floyd, like countless others who were forced out of work during the pandemic, sought escape in drugs and alcohol. In March of 2020, he was hospitalized due to an overdose.[29]

By May, the country's lockdown policies were in full swing. The nationwide unemployment rate stood at nearly 15%, the highest recorded rate since the Great Depression. Fifty-four million Americans had gone hungry, suicide rates were skyrocketing, and opioid abuse had never been higher.[30]

Prior to lockdowns, the U.S. unemployment rate stood at an historically low 3.5%. A study by economist Harvey Brenner in 1979 found a 1% increase in unemployment is associated with around 37,000 deaths. A National Bureau of Economic Research study similarly found that for every 1% increase in unemployment, opioid-related deaths rise by

3.6%. *You do the math.* While the country's leaders were "saving" the populace from a cold virus, they were killing them *en masse* through lockdown policies.[31]

So, after two months of artificially induced chaos, an unemployed, confused, and confined citizenry was on edge. Difficult to convey to readers who did not live through lockdowns but imminently familiar to those who did, but a sweltering anxiety seeped throughout the country. By late-May, the United States was a powder keg building with pressure, primed to explode.

While grainy surveillance camera footage taped Latasha Harlin's murder nineteen years ago, two smartphone cameras captured the events of May 25th, 2020, in crystal-clear 1080p. The democratization of high-definition video recording enabled the documentation of what has befallen hundreds, if not thousands of individuals at the hands of police authorities—but only made visible with the advent of handheld supercomputers.[32]

Vladimir Lenin infamously said, "There are decades where nothing happens; and there are weeks where decades happen."[33] The same could be said of minutes, even seconds. Military and law enforcement personnel know the phenomenon well, but also anyone who has experienced an instant plunge into chaos, when the calm of everyday life is broken ferociously and without warning—individuals thrown into making life-changing decisions within a matter of moments. In the case of May 25th, less than fifty-five minutes is all it took for the course of American history to change.

***

It is an overcast Monday evening in south Minneapolis. At the intersection of Chicago Avenue and East 38th Street, the convenience store and deli, Cup Foods, is open for business across from an abandoned gas station. Locals are concluding their fifty-ninth day under lockdown restrictions. Security footage reveals some residents walking around in violation of social distancing protocols and not wearing face masks. The hotline for Minneapolitans to report these criminals was open but it is unknown whether anyone fulfilled their civic duty to notify the authorities.[34]

Sometime around 7:30 PM, a forty-six-year-old George Floyd enters Cup Foods and purchases a pack of cigarettes with a twenty-dollar bill. At 7:57, the convenience store's two employees suspect the bill is counterfeit and approach Floyd who is parked across the street in a navy-blue Mercedes Benz SUV. Floyd occupies the driver's seat flanked by his friend Morries Hall, while his girlfriend Shawanda Hill sits in the backseat.

Hall became friends with Floyd after being introduced through a pastor but later became his drug dealer. In the ensuing court case, he invoked his Fifth Amendment rights against self-incrimination as methamphetamine pills were found in the vehicle. Hill, on the other hand, later testified that prior to the police's arrival, Floyd "fell asleep" in the car and that she struggled to wake him up.[35]

The shopkeepers confront the group outside their SUV about the counterfeit bill in total violation of social distancing rules. Floyd and Hall deny the accusation and the shopkeepers are turned away. Hopefully, no one contracted the cold virus as a result. At 8:01, the employees call emergency services to report the counterfeit bill. According to the 911 transcript, they describe Floyd as "awfully drunk" and "not in control of himself."

Roughly ten minutes later, the first police officers arrive on the scene. Officers Thomas Lane and J. Alexander Kueng approach the blue SUV and begin speaking to Floyd. Bodycam footage from Lane clearly shows an erratic and non-compliant suspect. Lane unholsters his firearm and orders Floyd to exit the vehicle. When he refuses, Lane yanks Floyd out of the SUV and handcuffs him. The officers lead Floyd across the street to their police car and attempt to place him inside the backseat. Rather than comply, Big Floyd collapses to the ground, claiming to be claustrophobic and unable to enter the vehicle. Not yet with any pressure on his neck, he also claims to have trouble breathing. Kueng removes a crack pipe from Floyd's pocket as he foams at the mouth.[36]

At 8:17, nine minutes into the arrest, officers Derek Chauvin and Tou Thao arrive at the scene. Interestingly enough, during his off-duty hours

Chauvin worked security at the same Latin nightclub that Floyd worked at. Both pulled shifts on Tuesday nights for the club's weekly dance competitions. It is unknown whether the two ever crossed paths.[37] Nevertheless, a worse cop for the situation could not have responded than Chauvin. The Minneapolis native possessed a proclivity toward employing neck restraints for extended periods of time, having conducted the maneuver on a 14-year-old boy for seventeen minutes three years earlier. With Floyd exhibiting breathing problems, Chauvin's preferred tactics created a recipe for disaster.[38]

By the time Chauvin and Thao approach the scene, their colleagues have managed to place Floyd inside the back of the squad car, but he continues to resist. Chauvin moves to the other side and forcibly drags Floyd onto the city street.

A *New York Times* crime scene analysis using security cam footage and two bystander recorded videos, employs phrases like "for reasons unknown" and "we don't know exactly why" to describe actions like Lane unholstering his weapon and Chauvin removing Floyd from the squad car. The whole tone of the video is a predictable criticism of "excessive police force" with ominous music playing in the background. [39] However, the NYT video was published before the officers' body cam footage was released to the public. Footage which makes it obvious that Floyd was highly resistant and acting erratically. The 6'6", 220lb, muscle-bound Floyd was clearly not going to remain in the backseat of the cop car, leaving Chauvin with practically no other choice but to restrain him to the ground.[40]

Floyd's autopsy report would later confirm the presence of fentanyl and methamphetamine in his system. Judging by his aggressiveness and frantic behavior, he was likely suffering from methamphetamine-induced delirium. The adrenaline surge from the meth combined with the stress of being arrested spiked Floyd's heart rate, precipitating his hysterics. The depressive effect of the fentanyl functioned to further disrupt his bodily and psychological functions.[41]

At 8:18, two bystanders begin videotaping with their smartphones. Floyd is now restrained on the ground by all four officers—Lane holding his legs, Kueng on his torso, Chauvin's knee on his neck while Thao stands guard against the growing crowd of onlookers. Floyd begins to call out "I can't breathe" and for his mother multiple times, yet Chauvin maintains the neck restraint.

Minneapolis police protocol stipulates that officers may only use a neck restraint if the suspect is "actively resisting." Video evidence clearly indicates that half-way through restraining Floyd on the ground, he stopped resisting. Thus, in maintaining the neck restraint, Chauvin violated Minneapolis Police Department protocol. In total, Chauvin kept his knee on Floyd's neck for nearly nine minutes. Floyd repeated sixteen times that he could not breath. He told officers "I'm about to die," to which Chauvin told him to "relax."[42]

At 8:25, Floyd's eyes close and he goes unconscious. A larger crowd forms on the sidewalk armed with more smartphone cameras, begging Chauvin to remove his knee. Suspecting Floyd may be suffering from delirium; Lane asks Chauvin to roll him onto his side but Chauvin refuses. Chauvin's eyes are cold and lifeless as he employs the neck restraint—almost as if he is not all there. Thao stands facing the crowd staring blankly as if he is thinking about something else. Thao later testified to merely serving as a "human traffic cone." Had he acted closer to a human police officer, he may have saved Floyd's life. Chauvin did not testify at his trial, never explaining why he maintained his knee on Floyd's neck for as long as he did. We will likely never know why Chauvin made that decision.[43]

At 8:27, emergency medical services arrive at the scene. Chauvin keeps administering the neck restraint until EMT personnel direct him to stop. He assists loading Floyd's lifeless body onto an ambulance. En route to the hospital, Floyd enters cardiac arrest and medics perform CPR to resuscitate him. Their efforts are futile. Floyd is pronounced dead at the Hennepin County Medical Center around 9:25 PM.[44]

***

Two autopsies were conducted. One by the Hennepin County coroner's office and another commissioned by Floyd's family. Both ruled his death a homicide but disagree as to the cause of death.

The Hennepin County chief medical examiner determined Floyd died from "cardiopulmonary arrest complicating law enforcement subdual, restraint, and neck compression." Meaning, police restraint caused his death, but rather than inducing asphyxiation—Floyd's heart stopped. The report further determined that methamphetamine and fentanyl intoxication likely contributed to his cardiac arrest, but ruled neck compression as its primary cause.[45]

The privately commissioned autopsy concluded asphyxiation induced by Chauvin's neck restraint killed Floyd. According to this report, Floyd did not die in the ambulance but on the ground with Chauvin's knee on his neck.[46]

Most importantly but rarely mentioned, is that a post-mortem nasal swab of Floyd tested positive for COVID-19! By simply using the same logic that hundreds of hospitals employed around the country, we can more accurately determine Floyd's *true* cause of death. He did not die from asphyxiation nor fentanyl intoxication nor cardiac arrest. George Perry Floyd Jr. was killed by COVID-19![47]

In similar fashion, Patricia Dowd, who according to CNN was "seemingly healthy," became the first confirmed death from COVID three months earlier. The logical deduction being that *"anyone can contract and die from COVID. This justifies every conceivable public health measure!"*[48]

What was omitted in any news reporting at the time, and only reported the following year once lockdowns began to subside, is that she had a massive heart attack! This was the coroner's ruling for her death! She just happened to also have contracted a cold virus. Nonetheless, "she died from COVID." This marks the beginning of the anti-scientific and destructive practice of categorizing deaths from completely unrelated causes to COVID—as what occurred with George Floyd.[49]

Just as medical professionals and experts witnessed patients suffering from signs of cancer or traumatic injuries from a car crash—but chose to rule their death from COVID—so did the obvious signs of neck compression and drug overdose deceive the public as to Floyd's true means of demise! Of the thousands of data points CNN and MSNBC gathered to broadcast on their telecasts to stoke fear in the population, Floyd undoubtedly contributed to the number of "new cases and deaths" from COVID in May 2020. Nonetheless, aerosolized particles are invisible to the human eye and millions who later watched footage of Floyd's arrest were deceived into believing nine minutes of neck compression precipitated his death.

As bystander videos of the incident percolated throughout the internet, a sweltering cauldron of racial animosity began to brew. Would Chauvin have retained the neck restraint had Floyd been a nineteen-year-old white girl? Would Chauvin have ignored the pleadings of a white man yelling "I can't breathe," telling him to "relax?" The footage seemed only to confirm what nearly two hundred and fifty years of real and perceived racial injustice suggested: America was irredeemably and systemically racist.

George Floyd joined a long list of names alongside Trayvon Martin, Eric Garner, Michael Brown, Freddie Gray, and Breonna Taylor—but also Emmett Till, Rodney King, and Latasha Harlins. All blacks whose murderers were either never charged or convicted. Thus, millions of unemployed citizens took to the streets to protest the latest manifestation of American injustice.

Initially peaceful, violent riots erupted in nearly every single major American city. Mass looting, arson, and clashes with police consumed the nation. Critically, the unemployment rate for young adults aged sixteen to twenty-four-years old exceeded 25% at the time. Take a wild guess at what age group comprised the majority of protestors.[50]

With most businesses closed and vacant, urban America became a practical free-for-all for looters to pillage and arsonists to burn. Two hundred municipal governments imposed curfews by early June in response.

Thirty-eight states activated their National Guard—its 96,000 deployed service members constituting the largest military operation other than war in U.S. history.[51]

In Minneapolis, rioters forced the evacuation of the Third Police Precinct and burned it to the ground. Floyd's Latin dance club was torched as well. Rioters inflicted $500 million in property damage with over fifty buildings flat-out destroyed. In response, Governor Walz dutifully told the public that protests were "no longer about the murder of George Floyd. It is about attacking civil society, instilling fear and disrupting our great cities." Thankfully, a sign language interpreter provided interpretation of his message for that sizable group of deaf people who lack closed captioning capabilities on their television. Nonetheless, Walz, who had shut down his state's economy and society, now advocated for its well-being.[52]

Similar scenes took place across the nation with rioters destroying over $2 billion worth of property. Governors and politicians responsible for creating the conditions necessary to spark the biggest riots in American history, sheepishly appeared on television demanding they stop. The President, who permitted epidemiologists and unelected bureaucrats to broadcast their prescriptions to lock down the country from the White House Briefing Room, deployed the 82nd Airborne and 3rd Infantry Division to quell the unrest. On May 29th, Trump tweeted that "when the looting starts, the shooting starts."[53]

The media dutifully played their role as well. Like magic, the most disruptive upheaval to world affairs since World War Two, drifted into the background of their news coverage. Somehow a cooped up and unemployed populace exploding into a fiery rage of fury and destruction had no connection to their forcible confinement. Media outlets characterized the riots as "mostly peaceful" and an understandable response to centuries of injustice. Suddenly, mass gatherings were no longer dangerous superspreader events but necessary exorcisms of America's racist past. The same virus that justified the shutdown of all society was no longer a concern.[54]

The riots would continue until lockdown restrictions were lifted. Walz was correct—the protests over police injustice had mutated into a spasm of violence untethered from the murder of George Floyd. What he failed to mention is that they spawned from his own policies, and those of leaders like himself. The protests did not end because "the message had been sent" against police misconduct or that racial injustice had been eradicated.

Rather, Walz and other leaders would have maintained the lockdowns indefinitely had they not faced the total evisceration of their tax bases. Put differently, the scale and intensity of the riots that summer forced public officials to ease lockdowns—not because the COVID-19 threat had diminished, but because maintaining strict restrictions in the face of widespread civil unrest became politically and socially untenable. 55

Two months. Roughly sixty days passed between the implementation of lockdowns and George Floyd's murder. Two months is all it took for our social order to crumble. A carefully calibrated performance crafted through millennia undone in a matter of weeks. Law and order eviscerated, revealing the strands that bind us together as thin and ephemeral. It turns out, you stop the system only briefly, and it all comes crashing down like a deck of cards.

"Systemic racism." "State-sanctioned oppression." "Police brutality." All mere disguises for the true means of breakdown. The true madness. Forcibly unemploying and shackling the population indoors was bound to end in a chaotic spasm of disorder. The population can only take so much before it cracks. The people can only take so much before they break.

MEDIA- HYPOCHONDRIATIC COMPLEX

"Fire one missile at the flag ship of each fleet. The Chinese will think
the British are rattling the saber. The British will think the Chinese are
being belligerent. And the media will provide cool objective coverage.
Let the mayhem begin."

— *Elliot Carver, Tomorrow Never Dies*[1]

The American media shoulders much of the blame for the COVID
debacle. How a simple cold virus mutated into a civilization-ending threat
stems from their deliberate manipulation—the nature of the fourth estate
being responsible. Journalists, or let us call them what they really are,
glorified observers, watch from the sidelines as real risk-takers sacrifice
and battle in the arena of life. Safe behind the ceramic rods of *The New
York Times* newsroom or beneath fluorescent lights at *Slate* or *Politico*,
professional onlookers impatiently contemplate how to make their
impact. Neither on the inside or outside of society, they occupy the
"media"—the Latin word for middle. Rather than venture onto the
battlefield of human endeavor, journalists sit comfortably "reporting"
what others do. Having studied "agenda-setting theory" or "framing
theory" at liberal bastions like Northwestern or Chapel Hill, bright-eyed
graduates yearn to "hold power accountable!" When in reality, they carry
intense biases—usually left-wing—and resolve to infuse those beliefs into
their work. Much like those overeager public health officials anxious to
justify their degrees, journalists pine to "save the world." If they cannot
influence directly, they will manipulate the perception of those who do.
Thus, while the press serves the societal utility of presenting the public a
rough idea of reality—for the most part, they function as an activist organ
for the disgruntled and resentful.

The psychopathy is only reinforced by implying one side is more
truthful than the other. Former President Barack Obama claimed
consumers of Fox News "live on a different planet" and credited their
inability to engage in constructive conversations to an erroneous

understanding of reality.[2] The deductive conclusion being that those who watch MSNBC or read *The New York Times* occupy the "real world." In truth, Fox News is by no means a reliable source of information any more than MSNBC, CNN or any other news outlet. Each has been reduced to a propaganda arm of a particular political interest. While the overwhelming ideological bent of American media leans leftward due to the inclinations of those who enter the profession, the fact remains that *all* news sources are tainted by political motivation.

Even self-proclaimed unbiased and neutral sources like the Associated Press and Reuters lean leftward. In 2024, a tweet jokingly claimed that Republican vice-presidential nominee J.D. Vance fornicated with a couch in his autobiography *Hillbilly Elegy*. The joke spread through social media with some beginning to believe it was true. In response, "fact-checkers" from the Associated Press combed through Vance's autobiography and confirmed there was no such story— publishing a rebuttal under the headline "No, J.D. Vance did not have sex with a couch."

However, later, as guardians of veracity, the Associated Press acknowledged it could not confirm with 100% certainty that Vance never in his life had sex with a couch. The logic being that Vance never explicitly admitted to humping furniture but maybe he did and kept it a secret. So they retracted the story, sparking a new round of speculation because the article was removed—setting the precedent that any accusation is unfalsifiable. Though, one suspects had a similar accusation been leveled against Kamala Harris or Tim Walz, the AP would have remained more circumspect about the nature of truth. Herein lies why Americans distrust the media. Even in an effort to conduct a public service by dispelling a classic case of misinformation—a joke mutating into hoax—the press corps failed to accomplish this basic task.[3]

The right remains no better. The Alex Joneses of the world poison public discourse in their own unique and schizophrenic way. As society grows increasingly complex, the attraction of attributing that complexity to faceless puppet masters pulling the strings on global politics intensifies. A bridge collapses in Baltimore? Conspiracy. A hurricane rolls through the Southeast? Conspiracy. An assassination attempt is made on a politician? Conspiracy. When two American soldiers were killed in a 2023 roll-over accident in Alaska, "truth seekers" declared they had been killed fighting in Ukraine. The idea being the Biden administration concocted the roll-over story to cover up their true cause of death. The tragic fatalities of two American servicemen co-opted to promote a depraved political attack.[4]

What is the penalty for this? Nothing. Just like when an athlete or coach "guarantees" a championship. If they do win, they're a legend. If not, everyone forgets about the guarantee anyways. Alex Jones can make one hundred outlandish claims, ninety-nine of which are total nonsense, but one will invariably prove half-true. Accusations that "all of Hollywood" or "every politician" engages in pedophilia or sexual depravity somehow justified by a single offender. Thus, declaring "Alex Jones was right" is rendered acceptable, licensing the whole cycle to begin again.[5]

Mainstream sources bemoan the corrosive effects of Alex Jones and other purveyors of "fake news" but fail to admit nor understand their own role in causing it. Is calling the President of the United States a Russian spy any less debased than denying the Sandy Hook shooting occurred?[6] Is denying that COVID-19 leaked from the Wuhan Institute of Virology any less dishonest than crying election fraud? Their insistence on labeling anything at odds with their preferred reality as "conspiracy theory" only fuels the Alex Jones-psychzo complex. For every obviously legitimate story the mainstream media dismisses, and for every obviously false one they promote, the public further and further drifts towards the seduction of charlatans.

The media's collective delusion stems from a fundamental misunderstanding of truth itself. The interconnecting powers of technology and illuminating prowess of science spawned an idea that deception and propaganda were antiquities of the past. In other words, the muckraking journalism of the eighteenth and seventeenth centuries, or the "yellow journalism" of William Randolph Hearst was behind us. Spreading lies was impossible in an age of instant and total access to information—or so the thinking went.[7]

Journalists quiver with excitement at the notion they possess some sacred line of truth and are the avowed guardians of this citadel of fact. In reality, the basic premise of being a "fact-checker" is arguably the most presumptuous and arrogant thought imaginable. To not only believe in some unitary, unwavering, and immutable set of truth, but to believe that *you* possess it, is the height of hubris.[8]

As has always been the case, truth is more a mosaic than one clear picture. For example, the Battle of Waterloo occurred in 1815 on a battlefield in modern Belgium. This is a fact. But what about the reasons why Napoleon was defeated? What were the battle's implications? Were British or Prussian forces primarily responsible for the victory? The "answers" to these questions masquerade as "facts" but are really just interpretive arguments supported to varying degrees by evidence. Ask an Afghani child whose father was killed by a U.S. air strike about the "truth" of American involvement in the Middle East. Ask another child whose village was liberated from the Taliban the same question.[9]

This is not to indulge in some kind of postmodernist refutation of truth itself. As described above, standards like beauty and excellence are objective. The 6th Arrondissement neighborhood of Paris is beautiful. Skid's Row in Los Angeles is not. When carbon is burned in the presence of oxygen, it *always* forms carbon dioxide. Outside of very high speeds and at the quantum level, Newton's laws of physics *always* ring true. However, in the realms of economics or politics or war—or public health—the casual factors to outcomes are so complex that "truth" bends and refracts. The game becomes who can leverage evidence in the most compelling way to persuade a majority of their audience. Nevertheless, STEM majors hailed the *Moneyball*, peer-reviewed, data revolution as the antidote to this "uncertainty"—that somehow a two-axed graph *always* trumped intuition and individual perception. *"Your own two eyes may tell you one thing, but you're wrong because this fancy looking graph says so. If you don't believe me, you're an anti-science imbecile."*[10]

Like any utopian fantasy, reality crushed this delusion. A research study can purport to "prove" such and such about reality, but another lies in wait ready to refute it. Evaluating the hypothesis that "socialism improves quality of life" yields different conclusions when comparing data from Scandinavia versus Venezuela. Analyzing temperature trends in the Arctic (which are rising) produces varying conclusions about man-made effects on climate change compared to measuring those in Europe and the United States (which are falling). The death rate of overweight, asthmatic COVID patients differed from their marathon-running, slim counterparts. Indeed, two seemingly contradictory claims can be true at the same time. Just because *some* individuals required lockdown measures does not mean *we all do.*[11]

In similar fashion, Trump advisor Kellyanne Conway infamously cited "alternative facts" to refute claims she was being untruthful. The media predictably pounced on the statement, spewing the same tired tropes of how CNN or *Slate* knew the "real facts." No doubt, Conway was vomiting out some Trumpian distortion of truth, but the media regularly engages in the same behavior.[12]

This is not criticism, but a statement of fact regarding the spatial limits of newspaper print. Billions upon billions of "stories" happen every day from a bomb going off in Yemen to a man cheating on his spouse. Editors pick and choose what to report and what not to report. This filtering of world events is distilled onto the pages of *The Wall Street Journal* or *Washington Post*, forming a picture of reality. However, those pages would look very different if every journalist reported from the South Side of Chicago as opposed to Beverly Hills County. A media outlet could report a district judge convicted a man on bribery charges but fail to mention the same judge is infamously corrupt himself. Therefore, citizens routinely label the world as "messed up" or "cruel" or "in decline," but are merely regurgitating their preferred media-filtered reality.

Furthermore, in the age of industrialized, mass-produced media (i.e. television, cinema, literature, etc.) truth has never been more malleable. As Oscar Wilde wrote, "Life imitates Art far more than Art imitates Life." If Wilde felt that way in 1889, with circuses, Vaudeville shows, and political cartoons practically the only forms of entertainment around, then one can only imagine the mimetic impact of art today.[13]

In the 2001 film *The Matrix*, before taking the red pill and escaping a simulated world controlled by robotic overlords, Neo lives as Thomas Anderson—a computer programmer by day and hacker by night. In its opening scenes, Anderson stores his hacked software inside a hollowed-out, but real-world book entitled *Simulacra and Simulation*.[14]

Far from arguing the belabored and faux-edgy point that "we live in a computer simulation," its author, French philosopher Jean Baudrillard, put forth something far more profound. Practically impenetrable to read due to issues of translation between French and English, but also the esoteric style of 1970s postmodern philosophy, the point is best explained through example.[15]

Imagine you are the first person to map a piece of terrain. Being the first cartographer, you have no preexisting maps to reference. You must walk the land yourself, noting every tree, every river, every hill visible to the human eye. You are an outstanding cartographer, meticulously surveying the terrain and depicting it unerringly on a map. However, no matter how good you are, no matter how detailed your map is, inaccuracies are inevitable: a stream you mistook for a river; a bush you thought was a tree; a mound you confused for a hill. The map, no matter how well-made, can never be 100% accurate.

Another aspiring cartographer, of lesser quality than yourself, wishes to make his own map of the terrain. But given yours already exists; he references it to help make his own. He corrects some of the errors—but not all of them—and makes new ones. That forbidding valley in the east is too dangerous to survey himself, so he copies your rendering. Why bother to chart the size of that swamp in the south, if your map already did so?

This process iterates again and again until the newest maps are markedly different than the actual terrain itself. They are "generally" similar to the source material (the terrain) but sufficiently distinct to where discrepancies become significant. An interesting phenomenon on its own but what happens next is truly profound. A traveler using one of those distorted maps plans to walk a depicted road. When he approaches where the road should be, he finds nothing but undeveloped land. A flaw introduced by the map-maker—caused by similar errors from previous cartographers.

Initial confusion turns to inspiration. The traveler vows to *create* the phantom road himself. The road is then paved to mirror the fallacious map. The encrusted errors of "simulating" the terrain have become the terrain itself! What is more real: the terrain or the map?

Better yet, let us take a more familiar example: the 1990s sit-com *Friends*. The show depicts living in New York City as white, mid-to-late twenty-year-olds in the 1990s. On the surface, its source material is the lived experience of white, mid-to-late twenty-year-olds in New York City in the 1990s. However, did the show's writers or actors conceptualize *Friends* in a vacuum—never having seen other literary, television or cinematic representations of that lived experience, or at minimum of something resembling it? Of course not. Living in modern, Western society entails being exposed to countless dramatic representations (simulations) of all facets to human existence—past, present and future. The "map" that is *Friends* was based on lived experience *but also* numerous other representations. Thus, the simulation is based on other simulations—which were in turn influenced by the same phenomenon.[16]

Finally, the behaviors, identities, and consciousness of both the audience and creators are subtly, or not so subtly, shaped by its consumption. A woman better understands what she wants from her relationship by watching Rachel and Ross. A man realizes his sarcastic wit is merely a coping mechanism for insecurity by watching Chandler. *"It's just a silly TV show. People don't shape their personalities around it!"* Not entirely, but to think it has no impact is clearly naïve. Children not only dress up as, but *believe* they are superheroes or princesses after watching Marvel and Disney movies. Do you think the phenomenon stops in adulthood? Thus, the source material—actual lived experience—is partly defined by its simulation.

In many respects, the same cowardly impulse that drives journalism fuels movie and television production. Having Tom Hanks and Matt Damon storm the beaches of Normandy functions to simulate bravery and heroism.[17] A corporate shill who works up the courage to confront his boss in one of the West's innumerable office dramas, licenses white-collar America an opportunity to cosplay their fantasies and exorcise resentments.[18] The immense popularity of *Game of Thrones* and Marvel films reflects a widespread desire to escape the monotony of everyday life, offering audiences a way to vicariously experience adventure and uncertainty. The appeal amongst affluent whites towards blacks rapping about inner-city life—despite the vast chasm between the two lived experiences—infuses danger into their highly predictable and tracked lives.

As Oscar Wilde wrote in 1889, the idea that reality mimics art is nothing new. However, its proliferation in the digital age is entirely unprecedented. What began with Gutenberg's movable type has evolved into a vast digital apparatus—a system through which simulations of reality now saturate modern life. Around 2,500 to 3,000 new movies are produced every year. The global total of new television content exceeds 100,000 hours annually.[19]

While a few avid playgoers in seventeenth-century England might have attended the Globe Theater regularly, a typical tradesman or nobleman watched a Shakespeare play once or twice a year. *Macbeth*, a simulation of medieval Scottish political intrigue—but more generally of the nature of power and ambition—surely influenced audience members in some respect.[20] Which in turn, influenced future artistic production. The demarcation between its source material and depiction was far clearer than it is today, however. That nobleman or tradesman returned home to a reality distinct and separate from *Macbeth,* thus further creating the "human drama" necessary to render future representation. Today, with screens oftentimes in every room of the house, and literally in our pockets, that line is non-existent. The simulation and its source material (reality) have converged. Far more interesting than the claim we inhabit a computer-generated world, but possibly of the same effect.

The end product of these overlapping representations piling on and melding into each other, which Baudrillard called the "hyperreal," creates something entirely new. The feedback loop between actual lived experience and its simulation so muddled and distorted as to form an undifferentiated mass.

As a result, audiences bemoan the prevalence of "remakes," claiming producers lack the ability to create original content. Granted, storytelling always relied on centuries-old, ancient archetypes and narratives (i.e. the Hero's Journey) but even exact characters and plots are now rehashed. Literal remakes are obvious to cite but even so-called "fresh" expressions are simple copies. Think Scorsese's *The Irishman* or any music by Greta Van Fleet or any *Star Wars* film produced after 1983. Human beings did not inexplicably lose their creative abilities in the twenty-first century. Rather, lack of source material itself is to blame, having been obliterated by the swamping effect of its boundless simulation. In short, there is nothing original to represent, except the representations themselves—with serious consequences for everyday life.

To illustrate the point, Baudrillard followed up *Simulacra and Simulation* with an essay entitled *The Gulf War Did Not Take Place* in 1991. Not some pathetic conspiracy theory denying the prosecution of the war itself, Baudrillard acknowledged that an armed conflict in the Persian Gulf where soldiers and civilians died did occur. Rather, he argued that the mediated experience of the conflict bore little resemblance to what actually transpired.

Any information passed through a media filter will always experience some degree of distortion—this is nothing new. We learn the twisting effects of sharing stories through a game of "Chinese whispers" as schoolchildren. However, the limitless nature of televised imagery mutated the phenomenon beyond measure. Through the innumerable prisms that translate armed combat, through its operational reports, then repackaged by journalists, and finally broadcast thousands of miles away to western audiences—the actual "Gulf War" vanished. A CNN news crew famously panned to a group of journalists in the region anticipating reports regarding the latest ground offensive—only to have them confess to watching CNN themselves to find out what was happening. The simulators had tripped into the simulation.[21]

In essence, Baudrillard argued images of F-117 Nighthawks, Abrams battle tanks, and A-10s created the perception that American forces were "fighting" Iraqi forces. In reality, the U.S. military so dramatically outclassed Saddam's army, in every conceivable way, that what transpired could hardly be called a "war." Armed with GPS, satellite imagery and night vision, American ground forces rolled over the Iraqi Army in less than one hundred hours. Instead, American operations functioned as a show of force to recalcitrant regimes of the consequences to disobeying the American-led new world order.[22]

Like all French postmodern philosophers, Baudrillard succumbs to that Marxist impulse to frame the conflict as some kind of capitalist-imperialist manifestation of deceit—that the moral failings of its prosecutors compelled them to "manipulate" and "deceive" the public.

If any regime is guilty of simulating reality for political gain it was the Marxist-inspired communist systems of the Soviet Union and Maoist China. Citizens living in the late Soviet Union coined the term "hypernormalization" to describe living under a system everyone knew was collapsing—but still pretended was working. The sheer scale of manipulation required to make millions of Chinese believe in backyard furnaces and fake grain yields defies comprehension. Therefore, stomaching the horrors of "from each according to their ability, to each according to their need" mandates just as much, if not more, manipulation as upholding any capitalist superstructure.[23] Additionally, what Baudrillard failed to acknowledge in his postmodern, Marxist haze of contradiction and moral grandstanding is that power talks—especially in the form of F-18 Hornets, Apache attack helicopters, and the 1st Armored Division. What Clausewitz wrote in the nineteenth century is no less true in the modern age: "war is an extension of politics by other means."[24] The United States had a political objective—expel Iraqi forces from Kuwait—and employed its overwhelming military power to fulfill it. Inferior Soviet-made tanks, subpar training, lack of night vision, and other weaknesses of the Iraqi Army are disadvantages the United States cannot be expected to rectify—nor apologize for. Don't invade sovereign countries and you can expect your ramshackle army not to be obliterated in three business days.[25]

However, Baudrillard was fundamentally correct. Admitting to the public that their military effort was more spectacle than war was politically unacceptable. As a result, Washington leveraged the power of televised imagery to create the perception they had learned the hard-fought lessons of Vietnam by executing a "clean war" against a rogue dictator. Furthermore, while Kuwait retained sovereignty as a result, nothing much else changed. By pursuing Saddam into Iraq and overthrowing him, the United States would have violated the same principles of self-governance and sovereignty it claimed to uphold. Therefore, Saddam maintained power and continued to persecute Kurdish and Shiite minorities in the conflict's aftermath. Finally, his continued presence laid the groundwork for the 2003 American invasion of Iraq—and all the misadventures that followed. In essence, the *"Gulf War Did Not Take Place."*

The Pandemic Did Not Take Place could just as easily be written today. Ninety-nine percent of the populace "experienced" COVID through either getting a week-long cold or knowing someone who did. The vast majority never even knew someone who died from the virus. Yes, COVID tragically killed people, and their loved ones will no doubt take offense at that claim. But for nearly the entire country, the actual pandemic—meaning patients hospitalized and killed by COVID—remained an abstraction.[26]

The virus mutated into a media object: a constant scroll of numbers, curves, and hashtags. Public response driven not by direct experience but mediated panic, policy theater, and algorithmic logic. Images of dancing nurses and doctors serving as mute testament to the illusory nature of the travesty.[27]

Moreover, COVID was never eradicated and continues to circulate amongst populations to this day. Therefore, the premise that the "pandemic" and its countermeasures mitigated the virus was pure simulation. Yet, the lockdowns, the fear, the forced alienation from fellow human beings, and all the other tyrannical controls emplaced on American citizens were viscerally experienced. The cosplay of an actual pandemic created the real conditions to subjugate and dislodge the freedom of millions of Americans.

***

If the pandemic constituted the simulatory apex of modern media, then its campaign to portray the country in deep decline and ultimately irredeemable stands a close second. Again, rooted in the sad resentment that they themselves could not compete; journalists scream about the unjust and corrupt foundations of the country. Most explain away this deceit by arguing the "bad stuff" simply drives viewership. They claim no one wants to read stories with headlines reporting "Everything is Actually Pretty Good" or "Nothing Bad Happened Today." This is true to an extent as humans are naturally drawn to potential threats—our proclivity to stare at car crashes as an example.[28] However, disparaging the United States more so reflects conscious or subconscious feelings of inadequacy on behalf of those who enter the journalistic profession. If they could not compete nor feel comfortable in the twenty-first century, then the whole system must be depraved.

The effect is to disproportionately report stories that illuminate the worst and most heinous aspects of society—no matter how miniscule a share of behavior they actually represent. Browsing *The Daily Mail* or *New York Post* would have readers convinced machete-wielding, watch-stealing, terrorists stalk the city streets of America. Alienated teenagers incessantly shoot up schools and husbands kill their wives in droves.[29]

The result: fear and animosity grip the nation. Will that staggering, drug-addicted vagrant you must inevitably cross paths with downtown lunge at you, ask for money, or both? When a fellow citizen even talks to a stranger in public, most assume they want a handout or will possibly hurt them. The average American bus or train is a scene of silence—its occupants careful to avoid eye contact while buried in their smartphones. [30]

Thus, the "image" of America is a declining empire, wracked by decadence and despair. Yet, if even the slightest credence is ascribed to statistical analyses of crime, poverty, and health in this country, then the complete opposite is true—just as a cursory glance at COVID-19 fatality rates dispelled the media-induced hysteria of the pandemic.[31]

This is not to accept a serene, fantastical version of modern America. Fate has denied us the seductive bliss that citizens of client states like

Luxembourg or Portugal feel every day. There is a reason why nearly everyone outside the United States can name its presidents, even its political parties, while the average American does not devote an iota of thought to theirs. This is not done out of ignorance, but because *they do not matter*—whereas the effects of American choices reverberate around the globe. Pseudo-intellectuals from abroad shroud their criticisms of the United States in the language of moral and cognitive superiority. In reality, their destinies are so intertwined with that of the United States, that they have no choice but to care.[32]

Thus, riding the tides of history bestows a unique responsibility to not only vigorously reflect, but change to the demands of time. There is a self-criticism to the American experiment that breeds discontent but also rejuvenation. Yet, the fashionable perception of it being in decay has devolved below constructive criticism into a debased repudiation of the experiment itself. Doing so discounts the immense achievement and marvel that is the United States. Any criticism of the country is like critiquing the hitting form of a Hall of Fame baseball player. Most have no idea what they are talking about and even if the criticism is warranted; the overall quality of the player vastly exceeds his every counterpart. Go to Bangladesh or Yemen and see how phenomenally the rest of the world is doing.

Furthermore, if the country is truly irredeemable then why do hundreds of thousands of young men and women continue to raise their right hand and swear an allegiance to potentially die in its service? Why do millions flock toward the country on ramshackle boats, cross barbed wire and risk their lives to live within its borders?[33]

Cliché to say but unflinchingly true, the greatest threat to the United States will always come from within. While Rome faced external threats, its ultimate collapse emanated from internal decay and corruption. The fall of the United States will derive from similar means. Most likely from a rejection of the experiment itself, induced by a manipulated picture of what it represents.[34]

Consequently, living in the modern world demands citizens trust their instincts and principles despite what any *New York Times* columnist says "science" or "the research" dictates. The best defense versus the twenty-first century's blitzkrieg of data and information is common sense. The instinct that wearing face masks to combat the spread of an aerosolized pathogen was inane and pointless, despite what the "experts" implored, was well-founded. The instinct that the virus was not worth destroying the country over, despite the pleas of journalists and public health officials, was justified. Obviously, there are limits to individual reasoning and we ignore the illuminations of empiricism to our detriment. A positive angiogram for kidney cancer does not care whether you feel you have kidney cancer or not. However, blindly accepting the supplications of the fourth estate or any claimant to expertise is a path paved to hell.

The pandemic proved the stakes of doing so are no less than the obliteration of basic freedoms and the American way of life. Through a curation of selected images and deceptive statistics, the media proved themselves imminently capable of wreaking havoc onto society—a miniscule subset of human affairs leveraged to produce nation-wide panic. The question going forward will be whether we fall for it again.

CROSSING THE RUBICON

"The die is cast."
— *Julius Caesar*

In 49 BC, Julius Caesar crossed the Rubicon River into Italy sparking a civil war between himself and the Roman senate, led by Pompey. At the Battle of Pharsalus the following year, Caesar's vastly outnumbered forces defeated Pompey's army, paving the way for his accession to dictator. He proceeded to stack the senate with his own supporters, seize control of the consulship and proconsulship (the executive offices), and declare himself Pontifex Maximus (leader of the pagan church).

Sensing the obvious that Rome's cherished republican institutions were under threat, a group of senators led by Gaius Cassius and Marcus Brutus assassinated Caesar on the Ides of March, 44 BC. Ironically, in Shakespeare's dramatic adaptation, Cassius and Brutus are portrayed as scheming, two-faced conspirators who betrayed the noble Caesar. When in fact, the two senators were trying to protect the Republic against the tyrannical ambitions of a power drunk warlord. Nevertheless, their efforts were in vain as Caesar's actions had permanently destroyed any viable return to representative sovereignty.[1]

While the Republic would not officially fall until Octavius became emperor in 27 BC, the "die had been cast" the day Caesar crossed the Rubicon. Nearly seven hundred years of popular rule ended, never to be restored again. Rome had died.[2]

Was the pandemic a "crossing the Rubicon" moment for the United States? Only time will tell but it certainly marked a turning point. As is often suggested, either by those who genuinely believe it or by those trying to sound macabre, the United States is in its "twilight stage." That we most closely resemble those last days of the Roman Republic—a once great body politic diminished by the inevitable cycles of history—is a commonly shared belief among the populace. After all, empires rise and fall; dynasties come and go; and the brutal corrosion of time erodes all before it.

Romantics argue the golden era of American excellence lied somewhere in the agrarian freedom of the revolutionary era; or in the rugged individualism of the frontier; or in the postwar, white picket fence tranquility of the 1950s. The thinking goes that a mob of screen addicted, possibly gender fluid, Gen Zers and millennials, untested by the great crucibles of war or true struggle, have neutered the American spirit— descending us into a nihilistic decadence untethered from religious piety or patriotic duty.[3]

Is it true? If the expression of societal values and beliefs in tangible form, otherwise known as "art," is any indicator, then absolutely. For those who claim art is subjective—you are wrong. A quick glance at Da Vinci's *Last Supper* alongside Cy Twombly's *Tiznit* dispels the notion immediately. So will listening to The Beatles' "Hey Jude" after playing Lil' Uzi Vert's "XO TOUR Llif3." Google the Nancy Pelosi Federal Building in San Francisco followed by the Duomo Cathedral in Florence. The differences in quality are obvious and any refutations to the contrary are disingenuous or flat-out ignorant.[4]

Modern American cultural expression consists chiefly of worn-out superhero tropes, a debased mutation of West African musical traditions, and carbon copy country rifts to appeal to Southern whites. Having Marvel superheroes win every battle, listening to faux thugs rhyme about taking drugs and having sex, or watching privileged fraternity bros cosplay the Old South does not rival the Renaissance in terms of art or cultural magnificence.

Postmodernists crow that art is better than ever and claim criticisms are rooted in social constructs like race, class, and gender.[5] Read the following excerpts of lyrics side-by-side and come to your own conclusions regarding the "subjectivity" of art:

"Hallelujah" by Leonard Cohen[6]
That David played, and it pleased the Lord
But you don't really care for music, do you?
It goes like this, the fourth, the fifth

The minor falls, the major lifts

The baffled king composing Hallelujah

"I Do It" by Big Sean[7]

I do it, I do it

I do it, yeah, like that, boy

Boy (B-I-G Sean Don, n****)

N****, fuck yo' time, n****

It does not take a Ph.D. in musical theory to perceive the qualitative difference. *"Only whites think that! America is racist!"* Wrong. It is not racist to say that "Fat Juicy & Wet" by Sexyy Red and Bruno Mars or Ice Spice's "Think U The Shit Fart" (yes, that is the song's name) are objectively bad. Just because you do not possess a certain racial background does not mean you cannot identify bad art when you see it.[8] Sadly, much of contemporary thought now blames broader political participation itself for these perversions. The apex of twenty-first-century democracy—a kind of *c'est la vie* libertarianism—hailed the wonders of an unshackled, emancipated populace free to pursue their own desires. The only condition being that your actions did not hurt others in the process. Previously demonized lifestyles like homosexuality and transgenderism became widely accepted, along with increased drug use. After all, consensual sex, changing your gender, and getting high only affects oneself, right? Historical constraints on these behaviors were deemed anachronistic and bigoted, rooted in Puritan religious fanaticism —whose credibility was obliterated by scientific advancement. Only bigots and anti-science imbeciles would ever oppose the full-fledged embrace of these practices, so said the intellectual vanguard.[9]

Indeed, shaming homosexuals for practicing their lifestyle was an ill-conceived social convention rooted in ignorance. While a "gay gene" has never been discovered, there is a strong correlation between homosexuality and the expression of specific genetic variants, implying sexual orientation is largely hereditary. Furthermore, naturalists have observed gay behavior in the animal kingdom. Therefore, the Supreme Court ruling that rendered gay marriage constitutional—*Obergefell v.*

Hodges—violated the religious underpinnings of marriage itself but did little to hurt the republic. After all, everyone has a right to be miserable.[10]

Yet, similar in how mere equality was insufficient consolation for race activists, revenge was the next order of business for LGBTQ zealots. The requirements of "being tolerant" morphed from detached acceptance into forcing Christian bakers to bake gay wedding cakes. Activists plastered pride flags around every major city and watching television became an exercise in homosexual exhibition. You would not just need to accept these practices; you would need them flaunted in your face.[11]

Finally, the odd, but seemingly non-threatening choice of an extremely tiny subset of the population to undergo mutilating, gender-altering surgeries was transformed from a peculiar subculture into the mainstream. This overreach from tolerance into the downright bizarre, demanded hulking, broad-shouldered and square-jawed androgynes wearing sun dresses be permitted to enter women's bathrooms. Accommodating the unnatural and disturbing became the end-all, be-all of American society. Somehow, the struggle for independence, the fight for broad-based political freedoms, and blood-spilling of thousands of soldiers were all meant to elevate the well-being and comfort of a gentilic-butchered microcaste of society.[12]

So, for reasons not fully comprehensible, the party of the working class and minorities—the Democratic Party—embraced the mantle of transgenders in the 2024 election. On the campaign trail, Democratic presidential nominee Kamala Harris said she supported incarcerated transgenders be provided access to gender-affirming care. President Joe Biden interpreted Title IX legislation—which was passed to protect women—as justifying transgender men participate in women's sports. Muscle-bound, XY chromosome-possessing, testosterone-fueled track runners and swimmers could now compete for gold against biological females.[13]

Much like how the country reacted to COVID, the reasons for these absurdities defy explanation. Rationality fails to explain why biological males would be permitted to compete athletically against biological females, let alone have that position incorporated into an electoral platform.

Is madness to blame? Ignorance? A kind of banal evil? On the face of it, its adherents will regurgitate something like *"it's about treating people with respect and dignity."*[14] But no matter your political persuasion, everyone draws a line of tolerance at some point. Society does not extend dignity and respect to murderers or pedophiles— extremists to the contrary who argue the latter should be sympathetically referred to as "Minor Attracted Persons" or MAPS (yes, they do exist). However, for the most part, society recognizes that individuals void certain privileges with their actions. The First Amendment protects freedom of religion, but a defendant accused of killing babies cannot justifiably argue his religion mandated he kill them. Any society, no matter how tolerant or permissive, has limits.[15]

So why would the idea that transgender men be allowed to compete against women in athletics, or use their bathrooms, or bend society backwards to accommodate them ever be taken seriously? Behind closed doors, proponents most likely relish witnessing just how much absurdity the country can endure; but its main catalysts lie elsewhere. Yes, the bored, suburban housewife looking for yet another "disadvantaged" cause to champion forms a large part of the phenomenon. Yet, ultimately, it is no coincidence that its manifestation has corresponded with a rise in mental illness.[16]

The dual developments of psyches believing they occupy the wrong gender and of those supporting their accommodation and exaltation are products of psychological disturbances. Only recently, as a result of the growing politicization of science, was gender confusion not considered a mental illness. For nearly fifty years, the Diagnostic and Statistical Manual of Mental Disorders (DSM) classified being transgender as a psychiatric disorder. Thus, by simply viewing today's proliferation of sex changes and their corresponding acceptance from the lens of only ten years ago, then the rate of mental illness has skyrocketed.[17]

The increasing visibility of inequality has induced its rise. Only recently have the "have-nots" of society been consistently exposed to the tremendous wealth and access of elites. A peasant three hundred years ago would have only seen a member of the aristocracy or, even more unlikely, of royalty, once in their lifetime.[18] Tilling the fields was not all that unpalatable if you never saw the inside of a palace. Now, the average person sees images of athletes and celebrities—the rich and famous— driving in sports cars, living in mansions, and obtaining access to large numbers of sexual partners on a daily basis. The inevitable feelings of inadequacy work to grind down the individual's psyche. In response, children and teenagers increasingly found that during the height of transgenderism, stating you occupied the wrong body was rewarded with attention and adulation. In other words, declaring your gender transition became a sure-fire way to catapult oneself up the social hierarchy.[19]

Like any social revolution, the hierarchy was flipped, turning transgenderism into a badge of honor. Social police wardens, aware that discriminating against homosexuals was firmly on the wrong side of history, repackaged skepticism toward transgenderism as merely another manifestation of bigotry and "hate." The letter "T" desperately hung onto the end of the acronym "LGBT" as a result.[20]

In reality, denying the right of gays and lesbians to embrace their genetic hardwiring is drastically different than expressing skepticism toward individuals making permanently life-altering decisions induced by mental illness—especially when children are involved. Nonetheless, the screeching crow of degenerative liberalism, whose subsistence depends solely on morally shaming its opponents, executed its search and destroy mission against anyone skeptical of the practice. No doubt, that intellectual current unrestricted leads to a kind of Mao's Red Guard of transgenders "correcting" the intolerant. A mob of 6'4", deep-voiced, androgynous beings in mini-skirts throwing "transphobics" off buildings was likely not the future the Founders envisioned.

Yet, their cause was the one Democrats chose to support to win battleground states like Pennsylvania and Michigan. In effect, the Democratic Party abandoned its working-class roots to embrace the anxious and mentally disturbed.[21]

***

Nonetheless, Democrats constituted the only political body capable of checking the subsequent ravaging of the Constitution. In abdicating their historical responsibility to safeguard the interests of disadvantaged citizens in favor of strange subcultures, the party held the door open for Donald Trump to return to the White House. Not only that, but any resistance that once curtailed the Right vanished, as if it never existed. The Great Capitulation, the Great Collapse, whatever you want to call it —but somehow the previously so loud and lecherous opposition, represented by the administrative state and media-industrial complex, simply gave up.[22]

It appears decades of obstinate resistance at anything the Right would do, no matter how logical or sensible, finally caught up to them. The cumulative effect of clamoring for lockdowns, fighting for men to use women's restrooms, violating the civil rights of particular races, and whatever other absurdity caught their attention in the moment, at last exhausted the Left's reserves. So when it mattered most—they gave up.

While the inane screeching of *The New York Times* and other liberal rags was always annoying, their silence is oddly terrifying. Much like Caesar two millennia ago, no one man has ever accumulated so much power and authority than Trump. Indeed, the checks and balances that once restrained any one person from assuming too much power are now seen as a hindrance—a counterproductive obstacle to be eviscerated rather than respected.[23]

As a result, massive fraud and waste inside the federal government was systematically purged by Trump and billionaire entrepreneur Elon Musk. Their steamrolling of checks and balances would be disdainful if the target itself was not erected by the same trampling of democratic norms. It seems destroying what torching the Constitution created requires returning the favor in equal measure.

Libertarians railed against obscene and wasteful spending for decades to no avail. Whenever any suggestion was made to cut expenditures, or God forbid balance the budget, a journalist could find someone somewhere who would endure an inconvenience as a result.[24] Such and such disabled brother's cousin's autistic toddler would somehow be hurt if we dare curtail government spending. A bureau could have a $40 billion budget, ninety-nine percent of which went to kickbacks and useless programs but yes—there were a few thousand dollars that went to helping disadvantaged people. Therefore, accusatory screams greeted any calls to limit spending levels, arguing the poor and deprived would be "gravely" hurt.[25]

Moreover, college educated males discovered that telling the cute Education Studies major that "we should cut funding to the Department of Education"—no matter how much that would actually boost learning outcomes—was not going to help their reproductive chances. Nor was the Harvard Kennedy School of Governance going to react too keenly at publishing a dissertation about the ability of free markets to solve societal issues. Thus, the entire enterprise was beyond reproach.[26]

As a result, no viable presidential candidate could campaign on limiting the massive inflows of taxpayer money into the federal government. Libertarians consistently received less than one percent of the vote in local and national elections. There were simply too many people within the country dependent on federal handouts to campaign on limiting those handouts.[27] Trump, intimately aware of this, never promised anything of the sort of government reform that him and Musk eventually undertook. Yes, Trump was always a lower tax-less regulation Reaganite, but he never promised the full-scale dismantling of the administrative state he eventually undertook.[28]

Therefore, in hindsight, tackling federal waste and abuse necessitated what it always would: a Trojan horse-style takeover followed by a blitzkrieg obliteration of federal agencies and programs via executive fiat. The egregious waste of taxpayer dollars on needless extravagances like newspaper subscriptions, or spending millions to house illegal immigrants in luxury New York City hotels, or $20 million to create Sesame Street programming in Iraq, finally ended as a result.[29]

Clearly, much of this kind of spending was a front for money-laundering or sweetheart deals to preferred contractors. The $1300 "coffee cups" in the Air Force or the $25 million for "gender programs" in Pakistan were always just ways to move money around. But nonetheless, by 2025 the swindle was exposed.[30]

For now, the benefits accrued are mostly positive. However, in overturning Rome's republican institutions, Caesar also did some "good." For decades, the Republic imported grain from Africa and Sicily to dole out to the needy. By the time Caesar assumed power, the program was rife with waste and graft, as even the wealthiest started receiving handouts. In response, Caesar reformed the program and ensured only impoverished citizens obtained the entitlement.[31]

Historians agree his most important reform was in adopting a 365-day calendar based on the solar cycle. Corrupt priests had abused the preceding date system, known as the Republican calendar, by arbitrarily elongating months to extend political appointments. In reforming the calendar, Caesar not only standardized timekeeping—a luxury modern societies now take for granted—but inhibited the corruption of vested elites. Named after him, the Julian calendar forms the foundation of modern time-keeping, since supplanted by the Gregorian calendar in the sixteenth century (which added leap years).[32]

While not as sweeping a reformation to time-keeping as implementing an entire new calendar, Trump could take advantage of his *de facto* dictatorial powers to abolish Daylight Savings Time (DST). Contrary to popular belief that DST was instituted to support the agricultural lobby, pushing forward clocks one hour in the summer is a holdover from energy conservation campaigns during the World Wars. Yet, research indicates the conservatory effects from DST are negligible and do not justify the ridiculous biannual ritual of changing our clocks.

Instead, the country should return to standard time all-year-long as the sun sets later in the summer due to solar patterns anyways. The maintenance of DST serves more as a reminder of the paralysis of federal government and its inability to make changes—no matter how obvious or sensible.[33]

The point is that much like how Trump arrived in a sclerotic and corrupt Washington, Caesar descended on a decadent and indulgent Rome. Assuming dictatorial powers enabled Caesar to institute much-needed reforms, but at what cost? History proves that given long enough, democracies devolve into paralysis where the needs of its constituents are abandoned in favor of serving insiders. True reform rendered seemingly impossible without overthrowing the checks on power itself.

The initial results are usually positive as when Caesar—and now Trump—each threw off the yoke of democratic restraint. However, in Rome's case, a "temporary" dictatorship led to the outright obliteration of the entire Republic. The subsequent Empire saw good emperors and bad emperors, but the inevitable result was gradual decline. The lesson of Rome is that dictatorship is only as good as its dictator. For every Augustus and Marcus Aurelius, there is a Nero and Caligula.[34]

***

So the United States finds itself at an inflection point all democracies eventually encounter. Does the slow, painful, and dysfunctional process of constitutional government justify its ostensibly limited gains? Its citizens look on enviously as East Asian autocracies build new bridges, hospitals, and airports in a fraction of the time it takes the West—and usually of higher quality. They see their cities dirtied and swamped by drug-addicted vagrants, while the streets of Shanghai and Beijing remain clean and safe.[35]

The promise of Western liberalism was a glorious future unshackled from "backwards" and "outdated" religious and parochial notions. Instead, American streets became filthy, its music and art perverted, mutilated men lumbering into women's restrooms made permissible, and a reversion to antebellum notions of race. American roads and cities decayed while supposedly "backwards" and "fascist" authoritarian regimes built magnificent high-tech rail systems and supercities.[36]

Granted they do not share the Western commitment to civil liberties like freedom of speech and due process. Indeed, the Chinese can build a bridge in less than a month, when it takes democracies ten years, but at least its citizens need not worry about going to jail for speaking their minds (modern Europe to the contrary). But the question has become: is it worth it? Or even more importantly: do American citizens still possess those same rights? Brick by brick the absurdities piled up, leading Americans to question the system more and more.[37]

For many, the COVID-19 pandemic represented that "last straw." Any lingering doubts about the virtues of shared government and its efficiencies were more or less confirmed. As local and state governments unemployed and interned their citizens, silenced dissent, and ripped away basic freedoms all in response to a cold virus with a 0.2% fatality rate— faith in the system plummeted.[38]

The social contract in China is that citizens sacrifice their political freedoms in exchange for iron-clad security and continuously rising standards of living. Whereas in the United States, citizens accept the waste of their taxpayer dollars and chaotic political system in exchange for the general freedom to live as they please. When even this agreement is violated, the populace begins to seriously question the bargain itself.

Austrian bodybuilder and *Kindergarten Cop* star Arnold Schwarzenegger told Americans "screw your freedom" in response to their skepticism toward mask mandates and social distancing protocols. Predictably, the press celebrated his remark, firmly establishing the new social contract of the United States: *Sit down. Shut up. Wear your mask. Listen to the movie star.*[39]

As a result, Americans stood by helplessly as the artifices of government overreach, snatched more and more authority, more and more a share of private production, and trampled over the clauses of the Constitution.

The apotheosis of this trampling, in blatant violation of Article 1, Section 1 of the Constitution, was relinquishment by Congress of its duty to make laws—its entire purpose! Overthrowing the king and taking up arms and Bunker Hill and Yorktown and *all of it* ensured that American representatives—that the people elect—cast their votes for specific laws, thus channeling the will of the polis. Instead, beginning in the mid-twentieth century, Congress passed the burden of law-making to unelected bureaucrats, thereby circumventing the entire impetus of the Republic.[40]

Now, Congress passes a sweeping law like the Make America Safe Act or the Clean the Environment Act (they always name them something unassailable like the "Patriot Act," despite how blatantly unpatriotic or draconian its effects may be).[41] The law will stipulate an edict like "enforce rules that make America awesome," passing it onto whatever bureau or agency happens to be in vogue. There, hundreds of thousands of federal employees "interpret" that overarching legislation to then enact innumerable "rules," which are essentially just new laws. The rationale being that congressional representatives are not knowledgeable enough to address the nuts and bolts of a task like "clean up Chesapeake Bay," requiring the "expertise" of environmentalists to hammer out the details. The effect: the environment remains polluted with the added kicker that American representative democracy is ruined in the process. The legislative onus having been removed from the legislative branch itself affects to remove the very concept of the Republic. The will of the people dispensed with, the locus of power subsumed by central government, and *voilà*: the administrative state is born![42]

Thus, as Trump assumes near total authority to dismantle this illegal superstructure, no one can seriously object. The ship sailed long ago to "care" about constitutional provisions or separation of power. In 2008, Ron Paul, a Libertarian but running as a Republican presidential candidate, advocated for eliminating the vast deluge of alphabet soup agencies at the federal level and return decision-making power to the American people. The result: he was laughed off the stage! Somehow the idea of unelected bureaucrats determining the fate of the country, much like how the British King's Privy Council operated during the colonial period, became part and parcel of the American experiment. So much so, that calls to end this madness and restore the country's basic principles were met with shame and laughter.[43]

Rather than trust individuals and communities to solve problems, resolving issues became the prerogative of central governance. Such and such agency "needs more funding" emerged as the catch-all prescription for any challenge.[44] The impact was not to empower any kind of special problem-solving committee but to siphon responsibility from citizens to the state. Challenges would not be confronted by the individual and their family, or even by local governments, but by federal power. In effect, most Americans cannot even name their local representatives. Not because they are ignorant, but because doing so *does not matter*. Their destinies so intertwined with federal government that monitoring local politics simply became a waste of time.[45]

In short, American citizens less than a century ago would never have contemplated turning to Washington to fix problems far more challenging than ones we demand the President or Congress fix today. Natural dips in the business cycle became the province of the state. Then the money supply. Then education. Then poverty. Then food prices. Even the well-being of foreign countries. The remedy to any deviation from perceived utopia shifted from human ingenuity, innovation, and perseverance to the discretion of the executive branch and unelected bureaucrats.

For example, when Hurricane Katrina rolled through New Orleans and killed residents who were told to evacuate but chose not to; who did the media blame? President Bush! Somehow it was the executive's responsibility to answer for people who refused to evacuate, as if Bush himself should have dragged New Orleanians from their homes.[46]

This impulse only snowballed until mitigating the spread of aerosolized particles became the duty of governors and presidents. The winner of two world wars, creator of the greatest economic engine in history, and symbol of freedom and liberty the world over, reduced to a sniveling ball of paranoid, hypochondriatic fear-mongers.

OUR FRIENDS ACROSS THE ATLANTIC

"Europe cannot stay united without the United States. There is no
moral center in Europe."
— *Senator Joseph R. Biden, speech to the United States Senate,
December 13, 1995*

There was a time when Americans believed they lived in the greatest
country in history. Mocked by Europeans and self-hating Americans, they
believed it because it was true. Ridicule it, deny it, deride it; the result is
all the same: the United States of America is the greatest country to ever
exist based on metrics of political participation, wealth generation, and
cultural production. The "poorest" members of American society,
meaning those who receive welfare, live in air-conditioned, internet-
connected, flatscreen TV and refrigerator-possessing abodes—enjoying a
standard of living the destitute a century ago could only dream of. The
combination of shared political participation, fortunate geography, strong
natural resource deposits, and culture steeped in liberty and initiative
blended to create a behemoth of national wealth production and dispersed
prosperity unprecedented in world history.[1]

Sure, political bodies like Old Regime France or Ancient Persia
accumulated vast amounts of wealth but its fruits remained concentrated
in the hands of a select few. Democracies and republics existed in the past
like Ancient Greece and Rome but never on the scale of the American
experiment. Indeed, the model for almost every Western regime today is
an imitation of the U.S. Constitution. Europeans can scoff at its existence
all they want but they themselves adopted practically all its principles.[2]

For nearly two centuries, while stuck in backwards, indentured
monarchies, Europeans could only watch from across the Atlantic as
Americans enjoyed representative government and civil liberty. Yes,
slavery still plagued the young nation but the same existed in Europe's
colonies too. Practically every European power owned vast quantities of
slaves in Africa, Asia and Central America well beyond Lincoln's

Emancipation Proclamation. European cultures being so racist they would not even "deign" to live alongside them, unlike the Americans. Moreover, besides from a brief and violent spasm of representative democracy in France, Europe remained stalled in monarchic forms of government for nearly one hundred and fifty years past 1776. Just as the French Revolution spelled existential doom for the kings and queens of Europe, the United States has always threatened the Old World by its mere existence.[3]

Look no further than the Civil War as evidence. In the darkest days of the American experiment, as it finally resolved the abhorrent contradiction of slavery—Europeans rejoiced. In the words of historian Don Doyle, their aristocracy was "absolutely gleeful in pronouncing the American debacle as proof that the entire experiment in popular government had failed. European government leaders welcomed the fragmentation of the ascendant American Republic." Britain and France both provided the South with arms and warships as a result, knowing the end of American democracy entailed the greatest boon to their monarchies since the beginning of hierarchy itself.[4]

Unfortunately for Europe (but fortunately for the human race), the South's inane decision to launch an offensive war at Gettysburg, rather than hunker down defensively, combined with the superior industrial capacity of the North, ruined any chance of Confederate independence. Thus, Union triumph forfeited Europe's only potential claim to moral high ground—the existence of slavery within U.S. borders—while remaining subject to the authority of a single ruler themselves.[5]

The obvious advantages of representative governance eventually compelled Europe to copy the American model, culminating in the *de facto* abolishment of monarchy after World War One. Despite historiographical confusion (as possibly the most misunderstood conflict in history), the First World War was not an "accidental blunder" caused by a domino effect of interconnected alliances—sparked by the assassination of a Habsburg duke. Rather it represented the royal houses' last-ditch effort to stem the tide of republican sentiment across the continent, exemplified by the Third French Republic. Why else did the Kaiser declare war on France who had nothing to do with the assassination of Franz Ferdinand nor Serbian nationalism? Recognizing their days were numbered, the Hohenzollern and Habsburg dynasties aimed to squash European democratization once and for all by launching the Great War. If not for America's entry into the conflict in 1917, the German Army—only forty miles outside of Paris—would likely have been successful in doing so.[6]

Thus, any semblance of democracy or individual freedom Europeans enjoy today (though rapidly eroding) is not only the product of American example but U.S. deliberate intervention. The story of Europe post-1776, is nothing more than a battle between dynastic power versus popular sovereignty sparked by American independence. The French support of rebels at Chesapeake Bay and Yorktown was driven by long-standing animosity between the Bourbon and Hanoverian dynasties—not some romantic sympathy for the cause of popular democracy. King Louis XVI and the entire French court anticipated rebel victory would produce a friendly American monarch, not a repudiation of the institution itself—a consequence of George Washington's voluntary relinquishment of power.

[7] Hence, Louis sowed the seeds to his own execution by supporting the cause, as it inspired Robespierre and the Jacobins a decade later. While a democracy across the Atlantic was barely tolerable to Europe's aristocracy, one on their front porch was certainly not. Nor could the Catholic Church stomach its most cherished satellite—the French monarchy—falling out its orbit. For centuries, the Pope referred to the French king as *Rex Christianissimus*, "The Most Christian King," a title with no parallel anywhere else on the Continent. France and the Catholic Church were one in the same. Thus, Pope Pius VI mobilized all of Europe to save it.[8]

The resulting "Reign of Terror" was not a product of Revolutionary decadence but a logical defense toward fighting the entire Continent. Faced with the armies of Europe on their borders and within them, the revolutionaries had no choice but to brutally crack down on internal dissent. Killing soldiers of the papal military force within France, known as the Catholic and Royal Army, comprised the majority of executions by the Jacobins. Thus, the French Revolution failed not from internal contradictions rendering it any less noble than its American counterpart, but from overwhelming resistance by Europe's royal class. French revolutionary ideals were practically indistinguishable from those of Hamilton, Jefferson, and Adams. Faced with similar circumstances, the Founders would have undoubtedly resorted to similar means to save the American experiment.[9]

None of this is historical "revisionism" nor some jingoist, rah-rah, screed about the virtues of liberty and freedom. Instead, it unwinds the dominant historical narrative of anti-American bias and European attempts to absolve itself of their authoritarian impulses. Just because *"There was slavery!"* does not give Europe any kind of moral high ground over the United States.[10] Just because *"The French supported the American Revolution!"* does not place American citizens forever in their debt.[11] Nor does claiming *"The French Revolution failed because it was morally corrupted!"* disprove the fact that the rest of Europe crushed its chances from the crib.[12]

Europe is not, and never has been, some kind of model of democratic virtue. In 1995, then senator and future president Joe Biden proclaimed in a speech to the Senate, that "there is no moral center in Europe" and they could not stay united "without the United States."[13] This is as true today as it was thirty years ago, or one hundred and fifty years ago. Left to their own devices, Europe has only proved capable of cannibalizing itself. The history of the Continent is nothing less than a never-ending series of dynastic and ethnic disputes, occasionally interrupted by American intervention.

So it is no surprise after nearly a century of receiving copious amounts of U.S. taxpayer money—as the Marshall Plan bled into the Cold War—that Europe got disrespectful. Despite the ability to utterly dominate Western Europe, much like the Soviet Union did in the east, the United States chose to rebuild their bombed-out and decimated society.[14] While thankful at first, gratitude turned into condescension. Whether in the form of snide laughter about American "materialism" from a Dutchman sipping his Starbucks while working on a MacBook, or outright repudiation within the highest echelons of European politics—the Old World had taken its relationship with the United States for granted.[15]

A cute quirk to be gently laughed off like most things the Europeans do, if not for the fact that their entire national security apparatus is dependent on American benevolence.[16] Unfortunately for them, the previously sufficient smile and chuckle as Americans sat secure in their wealth and abundance gradually lost its appeal. With their cities and suburbs crumbling, U.S. citizens naturally questioned why billions were sent to other countries, especially to those who disrespected them. Trump only stated the obvious by wondering aloud what exactly this return on "investment" really was. Incessant accusations of being "war-mongering racists?" Tariffs on American-made goods? Finally, Europe overstepped in trying to convince the American public that the Russo-Ukrainian War was anything but a brutal civil war between two corrupt regimes.[17]

Just as they did *twice* in the early-twentieth century (three times if you count deterring a postwar Soviet invasion), Europe begged once more for the United States to bail them out. The threat this time: Putin's Russia. Despite constant accusations of being the "world's police," Europe maintained it was America's responsibility to resolve their most recent militarized, ethnic dispute.[18]

A sympathetic Biden administration sent a cornucopia of howitzers, HIMARS, Bradleys, and practically any other military equipment necessary to wage modern war as a result. Indeed, enabling Ukrainian resistance did stem the recreation of the Soviet Union in Eastern Europe, as a lightning-fast decapitation of Kiev's government would have emboldened Putin to march into Moldova, Romania, and beyond.[19]

However, its cost was a razor-thin provocation of Putin deploying nuclear weapons. Bolstered by American-provided armament, Ukrainian forces were poised to burst through Russian defensive lines and potentially retake Crimea in late 2022. American intelligence received intercepts that Putin was seriously considering deployment of a tactical nuclear weapon in response.[20]

The operational advantages of nuclear weapons are overestimated but the psychological effect of their first use since 1945 would have been seismic. Ever since the Soviets gained nuclear parity in the late 1950s, *any* use of atomic weaponry has been a non-starter, despite the limited effects of low-yield tactical warheads.[21]

In the early Cold War era, the newly created United States Air Force correctly discerned receiving the bulk of defense budgeting depended on convincing policy-makers modern war hinged on an all-out atomic shoot-out against the Soviets. A persistent nuclear strike capability, centered on B-52 bombers and later bolstered by Titan and Minuteman missiles, ensured the Air Force dominated defense budgets. Compelled to build their own capability in response, the USSR turned the Cold War into a contest of who could deliver the most nuclear warheads on target in the shortest amount of time.

Rather than increase either's security, the idea functioned to ensure any use of nuclear weaponry, no matter how small, would trigger the employment of each side's entire arsenal—destroying the planet. Hardly a "victory" in any meaningful sense. The fact the world did not end in a fiery ball of nuclear conflagration serves as testament to the maneuverings of mid-twentieth century diplomats and doctrinal soundness of Mutually Assured Destruction (MAD).[22]

However, the strategic inertia of understanding *any* nuclear deployment as entailing the end of the world has held into the twenty-first century—despite fundamental changes to the nature of war itself. Long-range sensors plus the impossibility of achieving air supremacy due to advancements in anti-aircraft systems (as has been the case in Ukraine) has reduced peer-to-peer conflict to a stalemate. Low-yield tactical nukes could open gaps in defensive lines allowing armored breakthroughs once more. Yet, the stigma against their use—no matter how limited their geographic effect—precludes their employment. Napoleon Bonaparte revolutionized warfare in the nineteenth century by eschewing the sclerotic nature of siege warfare for the risky, but decisive effect of pitched field battle. The next "Napoleon" will likely be the commander capable of finally mastering the nuclear conundrum.[23]

For now, the grand strategic effect of the Russo-Ukrainian War is to convince Europe that their security is no longer guaranteed. Able to fund massive social welfare programs with relative ease given their defense was funded by American benevolence; Europeans would then turn and scoff at the lack of socialized medicine or education or whatever handout they could cite in the United States.[24]

For example, the British beam, almost orgasmically, at the thought of their cherished National Health Service (a close second being the prospect of referencing American school shootings), despite significantly poorer quality of care, shortages of medical equipment, and nauseously long wait times to simply see a doctor. In essence, the NHS is the biggest sacred cow in the entire kingdom. Operating under the delusion that the average American is kicked out of an emergency room for failing to flash a seven-figure bank balance, the Brits relish waiting months to receive "free" but subpar healthcare.[25]

The obsession is a Freudian manifestation of guilt from upholding an aristocratic landed gentry for centuries. "Social mobility" in Britain has always meant nothing more than being born into the right family, attending Eton, then Oxford or Cambridge, then serving as a backbencher at Westminster. There are no British Elon Musks, Steve Jobses, or Andrew Carnegies for good reason. As a result, the NHS is their *mea culpa* for perpetuating a totally stultified social hierarchy.[26]

What they neglect to acknowledge is that American funding of their national defense is what makes it possible—with the supposed "socialist utopias" of Scandinavia being no different. Europeans doggedly cite their welfare systems as evidence of some kind of moral high ground over the United States. When in reality, those systems depend on an economization of artillery shells and missile defense systems, because the materialist Americans foot the bill. Having alienated their benefactors, expect the sunshine and rainbow welfare states of Europe to quickly lose muster.

The French, led by Emmaneul Macron, are leading the charge in remilitarizing Europe to combat Russia. In another dose of irony, Russia is the only reason the French state still exists in its current form. After their surrender to Nazi Germany in 1940, French Prime Minister Philippe Pétain agreed to collaborate with Hitler's murderous regime, infamously declaring "l'honneur a été sauvé!" (honor has been saved!) following the armistice. Roughly 10-20% of the French population actively collaborated by informing on fellow citizens and passing intelligence onto the German military. Pétain deported nearly eighty thousand French Jews to Auschwitz. French soldiers fought and killed Allied troops in North Africa, Syria, Lebanon, Madagascar, Senegal, and Algeria. This explains why France so fervently emphasizes the role of La Résistance, despite resistors comprising less than 1% of the population. If not, the record of France in World War Two is a dark and dreary one, on par with Fascist Italy or Quisling-led Norway.[27]

Nonetheless, despite their minimal contribution and outright opposition, France wanted to share in the spoils of war following Allied victory, demanding a slice of West Germany. Churchill and FDR were furious, questioning why the collaborators would receive a share—let alone be allowed to retain sovereignty. However, the looming threat of Soviet aggression necessitated Britain and the United States maintain a strong and independent French state. Restoring French dignity became simple, Cold War strategic necessity. Thus, Macron can thank Putin's old employer, the USSR, for his country's current position.[28]

***

For now, the fun and amusing exercise of bashing Europe must end. Indeed, their cultural achievements are unparalleled. The intellectual foundation of the United States is merely an offshoot of those European ideals articulated during the Renaissance and Enlightenment. Moreover, their commitment to healthier lifestyles should be the envy of the world.

While the American food supply descended into ultra-processed glop bordering on dog food, Europe ensured the safety of its own. Differences in the ingredient composition of a Domino's pizza in the United States versus one in Belgium or Spain are striking. One looks like a mad science project gone wrong (as most ingredient lists of American food products do) while the other is relatively tame. American baby formula contains grotesque amounts of added sugars, a practice with no parallel across the Atlantic. Food dyes known to cause neurological disorders saturate the U.S. food supply while Europe banned their use decades ago. Libertarians can mock European overregulation and their stagnating economies all

they want, but at least their populations are not poisoned on a daily basis.
29

That being said, while American leaders are most responsible for the country's heinous response to COVID-19, much of its impetus originated in Europe. Outside of China, Europeans led the charge in determining just how draconian governments could react. Italian citizens required literal permits to leave their house. Spain banned outdoor exercise and issued fines up to €600,000 for violations. The French forbid travel beyond 1 km of citizens' homes. If you wanted to leave your house in Greece, you needed to text the government—stating your reason and await approval. Would Socrates and Aristotle have approved? A Pew Research poll found *every single* European populace rated the United States' response to coronavirus *the worst* of any country in the world. It appears anything short of total economic and social strangulation of the American people would have disappointed our "friends" in Europe.[30]

Of course, the epicenter of pandemic alarmism was the World Health Organization, headquartered in Switzerland. Any semblance that combatting a cold virus necessitated a calm and measured approach was immediately denounced and relentlessly attacked by WHO officials.[31]

Put differently, a kind of panicky malaise infects the Continent, and it is no coincidence that calls for lockdowns and social distancing permeated from within its borders. European culture is marked by a mind-numbingly sclerotic insistence on rules and regulations. As an example, during the planning phase of D-Day operations, the British infamously conducted overly cautious and deliberative staff meetings while the Americans simply decided upon a course of action and executed.[32]

Thus, as two hundred and fifty years of history has proven, Europeans will always present a chastising lecture to combat anything the United States does or stands for. The two cultures are entirely different, however, possibly to an even larger degree than between the U.S and China. The future depends on American citizens embracing their fundamental values and principles despite what any jealous or insecure European may say.

CONCLUSION

"The strong do what they can, and the weak suffer what they must."
— *Thucydides, History of the Peloponnesian War*

As a historian, I am not in the business of making predictions. That being said, some things are becoming so intimately clear they must be addressed. The postwar, shared security world order that dominated political thinking since the mid-twentieth century has ended—accelerated by the absurdities of the COVID-19 pandemic. Even before 2020, the virtues of collective security, free trade, and political correctness had grown exceedingly stale. Forever wars in the Middle East, financial system implosions, the degradation of art and culture, and other follies had proved that whatever the governing system was—it wasn't working.

Mistakes are inevitable and expecting leaders to manage the complexities of modernity to perfection is unreasonable.

However, when mistakes are made, honest reflection must follow, and if necessary, individuals held accountable. Yet, Americans have become accustomed to never having the perpetrators of even criminally negligent acts face consequences.

For example, despite deliberate manipulations to the nation's housing market plus legendary amounts of fraud preceding the 2008 financial crisis, only one individual was ever convicted of a crime. Kareem Serageldin, a Credit Suisse executive, pleaded guilty to manipulating the value of mortgage-backed securities—a crime nearly every single bank in the country did *en masse*. Furthermore, the insistence of federal government agencies—Freddie Mae and Freddie Mac—to guarantee home loans, no matter how risky or ill-advised, functioned to turbocharge the risk banks were willing to take. The Securities and Exchange Commission dished out some fines, but the primary "remedy" was multi-billion-dollar bailouts of taxpayer and Federal Reserve-printed cash to save the very architects of destruction.[1]

So a similar phenomenon applies and will apply to those who perpetrated the COVID-19 fiasco. One would think individuals who literally funded the development of contagious viruses would face some kind of punishment. Though, not a single researcher, NIH grant distributor, or gain-of-function zealot has ever even been indicted—let alone convicted. Any chance that some kind of justice would be served for Anthony Fauci disappeared in early 2025, when outgoing President Joe Biden issued a full pardon for the good doctor. Biden further sanctified this country's culture of "accountability" by issuing a *ten-year* pardon for his son Hunter. Meaning any potential crimes Hunter committed within a ten-year period are off-limits to the justice system. At this point, why not make it his whole life?[2]

Thus, for those above a certain level of authority, a *quid pro quo* system of favors and unspoken agreements replaces true accountability. Quick to "forgive and forget" others' wrongdoing, these individuals assume they will be granted the same leniency when their own recklessness comes to light. This "wink and a nod" system of justice at the highest echelons of American politics only serves to consecrate the conditions for future misdeeds. As a result, the system lumbers on—most likely to the next calamity.

The cumulative effect of these usurpations has been to reject the system altogether. The 2016 election of Donald Trump representing the first salvo of resistance towards its obsolescence. Late-night talk show hosts demeaning legitimately held beliefs, having jobs shipped overseas, witnessing the importation of millions of foreigners possessing cultures antithetical to American values, and other infringements forced the public to elect a boorish and deeply unpleasant New York City confidence man. The conniptic fits of the media and Left in response to his mere presence served to confirm the public's resentments. If there was any room for doubt regarding the incompetency of the establishment, their inane reaction to aerosolized particles floating through the air obliterated its last vestiges.

Some have characterized these developments as the death throes of a dying empire. Unable to come to grips with their reduced standing in the world, Americans have resorted to electing demagogues and desperate allusions to a bygone era. However, in terms of economic production, military prowess, and technological innovation nothing could be further from the truth.[3]

The United States remains the largest economy in the world despite ridiculous arguments that on the basis of purchasing power parity (PPP), China merits that distinction. Another jiu-jitsu move by the economics profession in attempt to manipulate reality. By artificially devaluing their currency, the Chinese Communist Party has made its goods cheaper, thus creating the illusion of increased prosperity.[4] Yes, $5 gets you further in Xianyang compared to Chicago but it says nothing of global economic influence or actual earning power. Furthermore, the Chinese economic model since the early 1980s has been to simply copy American innovation. Just as DeepSeek is a copy of ChatGPT; so is a J-20 fighter jet to an F-22, or Weibo to Twitter, or Huawei to Apple. The Wuhan Institute of Virology and its commitment to gain-of-function research only copied what the National Institutes of Health pioneered. In other words, the Chinese state capitalist model has proven itself capable of inducing tremendous economic growth but nearly no innovation.[5]

Therefore, the phenomenon of "American decline" is a deliberate choice, not preordained destiny. Allowing the country to be pillaged from abroad and within was a deliberate choice. Allowing out-of-touch experts to run society into the ground was a deliberate choice. Allowing race and gender to dominate the socio-political underpinnings of the country was a deliberate choice. Reversing its effects will be consciously done as well.

In many respects, that decision has already been made. The second election of Trump and collapse of resistance towards his authority represents a complete repudiation towards decades of mismanagement. Unable to influence the direction of the country itself, citizens settled for sending a *de facto* middle finger to the establishment in casting their vote for Trump. The concern is that checks and balances have become virtually non-existent during his presidency. The privilege of curtailing power relinquished by those who fought for race-based hierarchies, transgender dystopias, and the utter waste and theft of American taxpayer dollars. A new paradigm based on raw power, transactionalism, and brute force has emerged in its place.

What is its effect? It has become fashionable to discuss a new world order ruled by the United States, China and Russia—each claiming their own geographic sphere of influence. The United States reinstates Monroe Doctrine-esque control over North and Central America. The Chinese mimic a kind of Japanese-style East Asian Co-Prosperity Sphere in the Pacific, and Russia retakes ownership over the old Soviet bloc. The predicable response has been to denounce this arrangement as an anachronistic spasm of imperialism and immorality. In truth, suggesting more powerful entities dictate world affairs merely reemphasizes and states aloud the forces that have dominated human affairs since the dawn of time.[6]

The ancient world operated under the very same conditions. In 416 BC, the Athenian Empire, the predominant power of its day, battled Sparta during the Peloponnesian War. Seeking to establish a foothold on the ostensibly neutral, but strategically important island of Melos, Athens sought to counter Spartan influence in the Aegean Sea. Behind closed doors, Athenian diplomats gave the Melians an ultimatum: surrender and pay tribute—or be destroyed. Melian leaders accused Athens of acting immorally, arguing their neutrality entitled them to be left alone. Athens replied with the brutal, but timeless message that "the strong do what they can and the weak suffer what they must." Despite this warning, the noble Melians refused to relent. As a result, Athenian triremes descended on the island, executing the men and enslaving its women and children. In essence, Athens did what it could, and Melos suffered what it had to.[7]

Known as the "Melian Dialogues," chronicled by Greek historian Thucydides, the Athenians' message is no less relevant now than it was two and half millennia ago. Obviously, modern society has progressed beyond executing and enslaving the conquered, but the principles of dominance and coercion remain unchanged. Despite efforts to indoctrinate children in the idea that "physical violence is never the answer" and plastering of behavioral-reinforcement posters in classrooms preaching the virtues of "just be a good person," social conditioning cannot change what evolution demands.[8]

Humans live in hierarchies. While a classless world where all live in peace and solidarity is undoubtedly preferable, it is impossible. This is unchanging and will remain so until the last thermonuclear weapons obliterate all semblance of life on Earth. Our closest relatives—chimpanzees and bonobos—exhibit hierarchical forms of social organization. Archeological evidence indicates humans appointed leaders to manage crop yields and direct farming resources starting in 10,000 BC. Even neurological studies have confirmed specific areas of the brain like the prefrontal cortex and amygdala evolved to identify social cues associated with dominance and leadership. Moreover, higher levels of hormones like testosterone and serotonin are associated with higher social status.[9]

In modern society, the same attributes that determined status in the Paleolithic and Neolithic Ages are relatively unchanged today. Physical strength and fighting ability being discounted in modernity—though not as much as some want to believe. It remains unflinchingly true that hierarchical position remains determined by factors like physical prowess, looks (especially in females), cognitive intelligence, technical skills, and even deception and manipulation.[10] What certainly does not determine position is "being nice" or even kindness (at least how it is commonly understood). Indeed, the threat or exercise of physical violence *still matters*, especially when it comes to mate selection. We need not waste time in citing the cornucopia of behavioral studies proving the link between female sexual preference and certain male behaviors. It is sufficient to say they do not include "niceness" and "passivity."[11]

Nonetheless, these traits were inculcated into the nation's youth, most fervently into young males. Late twentieth-century children were raised to believe that the "bullies" of adolescence would fade away in adult life, only to discover they remained all too common into adulthood. Instead of tormenting you in the hallway, they were your boss—and they could end your career.[12]

Just as this indoctrination campaign swarmed through American schools, a similar idea took root in global politics. In essence, the legacy of the past two generations has been to absolve the world of its Melian dictate—that somehow social relations could be governed by kindness and a kind of kumbaya social construct.

In reality, the prescription merely disguised the same ambition to power and dominance. Model UN and Peace Corps zealots relished in the moral superiority of upholding values like "human rights" and "global justice." Siphoning off decision-making power from local and state governments to globalized institutions—which they were in charge of— was framed as "enlightened" and "progressive." Good and evil became binary, and obviously, you can guess who claimed to occupy the role of "good." This moral grandstanding was effective as who could possibly oppose the advancement of human rights and equity.[13]

Of course, none of these enlightened saviors ever contemplated entering socially useful, but physically demanding professions like construction, plumbing or the military. Instead, the managerial class could "do so much more" in fields like international affairs and political science. Instagram photos at the UN Headquarters in New York, or at the opening of a new water well in Zimbabwe served to sanctify the moral superiority of this awakened class.[14]

Yet, its only effect was to elevate the incompetent and unqualified. Banquet dinners in Davos and Brussels did not render globalist institutions any less ineffective. Rather than renegotiate the nature of world politics, power was merely transferred from individual nation-states to centralized, international governments. Any resistance to this takeover was framed in ethical terms. Moralists claimed nationalism and patriotism were refuges of bigotry and racism. Questioning the prescriptions of institutions like the World Health Organization or United Nations was ignorant, anti-science and yes—racist. Populations conditioned for decades to understand any expression of "racism" as the worst sin imaginable dutifully sheathed their concerns when confronted with the accusation. Its consequence was to cede authority to unelected bureaucrats lecturing from UN council chambers. Nonetheless, after decades of overwhelming evidence suggesting this was a bad idea, much of the world rejected their ideas completely.[15]

The implications of the new world order are yet to be seen. Certainly, Europe looks to be on the losing end of any power reshuffling led by the United States, China and Russia—with second-rate powers like Saudi Arabia and India running close behind. The ultimate losers are the weak and cowardly who somehow managed to co-opt the levers of political and cultural power for the past quarter century. Their age is over. Their crowning moment of stupidity being the COVID-19 pandemic.

Do not assume the Left or administrative state or whatever label you want to ascribe to the entity that formed the media-hypochondriatic complex is finished. When up against the wall, this group tends to lash out even more atrociously and nonsensically. In many respects, their reaction to COVID embodied a last-ditch effort to sink Trump's election chances in 2020. It worked beautifully! No one, not even a New York real estate tycoon could bluster their way out of aerosolized particles spreading throughout society.[16]

Nonetheless, the Left has never been more demoralized or diminished in its wake. Fatigued from its bouts with the indefatigable Trump and somewhat resigned to the reality that many of their ideas were flat-out stupid, they have rolled over completely. This development should not be celebrated as many on the Right are keen to do. The collapse of any viable alternative to the Republican Party only supercharges the nation's descent into demagoguery and authoritarianism. Thankfully, smarter minds in the Democratic Party are verbalizing the obvious that framing transgender rights as a new civil rights movement will not earn the votes of fifty-one percent of the electorate.[17]

Rather, a leftward turn economically will capitalize on resentments about wealth inequality. Any electorate is vulnerable to calls for wealth redistribution. Promises of checks in the mail function as a sure-fire way to win millions of votes. The Left would be wise to find a Bill Clinton or Barack Obama-esque candidate, but with the talking points of Bernie Sanders. If history is any guide, no matter how much destruction, pain, and suffering they inflict, evidenced by the tens of millions dead in Communist China and the Soviet Union—egalitarian doctrines always rise again, ready to pounce on the indignations of the inevitable losers in the capitalist game.[18]

***

Nor will a rising technological utopianism solve the challenges of modernity. Sam Harris, who advocated for the deaths of children to stimulate COVID hysteria, claimed that AI will progress to such an extent, whereby for every "week" it operates, it will advance the human race "20,000 years." Thus, every "six months," it will advance the world "500,000 years."[19]

Whatever that nonsense actually means may produce a provocative Ted Talk or garner news headlines, but is so removed from reality it boggles the mind. 500,000 years ago, proto-humans had yet to develop language, only recently began using stone tools, and struggled to maintain a campfire. Today, we have split the atom, landed on the Moon, and manufactured microchips capable of processing trillions of operations per second. Expecting comparable leaps every six months is not only fanciful, but a deranged overestimation of technology.[20]

Similarly delusional, futurist Ray Kurzweil predicted computer intelligence will surpass human intelligence, sparking a never-ending increase in the former—known as the "singularity." Never mind that everything in the physical world, from human height to the speed of light, has natural limits. There are no twenty foot tall nor 1000 IQ-possessing humans. Even Moore's Law, the oft-cited theory that computing power doubles every two years, has precipitously slowed down since 2016. Expecting that computer intelligence, unlike everything else in the known universe, will not reach a limit is beyond absurd.[21]

Put differently, while contrary to the wishes of STEM majors and Silicon Valley delusionists, the world is not a sum of its blueprints and engineering prowess. Their prescription for creating economic opportunities for the sons of West Virginian coal miners was to "learn how to code." As that predictably failed, they resorted to even grander delusions about artificial intelligence. However, AI will not produce a labor-less future where humans clock into Dopamine Maxxer 9000s to live out their days in sublime bliss. Robotic infantrymen will not march into battle and remove the necessity of human sacrifice to warfare. No matter how much a Ray Kurzweil or Sam Harris may push their fantastical predictions, humanity remains shackled to its eternal burden.[22]

Likewise, on April 2nd, 2025, Elon Musk tweeted that "it increasingly appears that humanity is a biological bootloader for digital superintelligence."[23] This encapsulates the thinking of the technological elite, who have unfortunately co-opted significant levels of power and influence in modern society. Their goal is not to serve humanity but exploit it at the altar of a debased and ignorant utopianism. Humanity is an end in itself, not a means to create something blatantly anti-human. Thus, society must guard against the designs of those who do not uphold mankind's best interests.[24]

The same applies to the Shark Tank-LinkedIn class. It became fashionable in the past quarter century to idolize entrepreneurs within American society. The pinnacle of human achievement being a kind of Steve Jobs-Mark Cuban career path. The zeitgeist assumed Business majors dutifully responding to the market's invisible hand would create socially useful products and push the country forward. Instead, we ended up with more brands of sugar water and Instagram Reels. Television audiences flocked to faux business shows like Shark Tank, to "learn" about venture capital and celebrate the triumph of entrepreneurialism. Yet, the best-selling product to ever emerge from the show was a sponge. A literal sponge shaped like a happy face. This was the crowning achievement of the entrepreneurial class.[25]

Getting rich has always been part of the American dream but never in such a slavishly narcissistic and graceless way. Bourgeoisie-aspirants flocked to study "Business" and "Finance" under professors who never had success in either field. Somehow the secrets to financial success could be "learned" from academics at the Stern School of Business and Wharton School of Finance. Graduates proceeded to pad their LinkedIn resumes with hollow, but sparkling claims to expertise or experience— their subsequent ascent up the corporate ladder framed as the noble product of hard work.[26]

In reality, most merely benefited from possessing above average IQs. Social scientists Charles Murray and Richard Herrnstein have demonstrated the overwhelming evidence linking higher IQs to higher incomes and social status. The tracked lives of America's cognitive elite pushed them into more lucrative professions not by any moral virtue, but by having won the genetic lottery.[27]

Lastly, emboldened by their financial success, this group attempted to throw their intellectual hat into the ring of geopolitics, history and practically any other field. In doing so, they not only embarrassed themselves, but inflicted disastrous effects in the process. Its culmination was the election of "businessman" Donald Trump and the embrace of a kind of America Incorporated. That somehow the country's problems were products of "bad deal-making" and unbalanced account ledgers. All nuance obliterated in favor of increasing profits and "making deals."[28]

Rather, there is a set of perceptions that undergird society, which shape its conditions. Spoiler Alert: they are not shaped who those who know Python or C++. For the STEM mind, reality fallaciously appears as a set of immutable "facts." Just as there is one right answer to calculate how much load a bridge can support depending on its length, engineers view the rest of human endeavor in similar terms. This explains why many members of those fields struggle with social interaction. The idea that truth is debatable in realms outside bridge design and computational programming confronts their minds like a computer error and they inevitably "crash." In actuality, the notions that structure reality outside of mathematical equations are guarded by gatekeepers known as historians, novelists, and poets. Thus, every time someone confidently proclaims a fact about the past, somewhere an historian smiles.[29]

Therefore, it is no coincidence that an upending of the world order is occurring simultaneously with a reevaluation of its foundational narratives. Currently, amateurs and professional historians alike are reassessing the most formative event of the last century—World War Two. The meta-narrative that the good guys beat the bad guys is increasingly under scrutiny—something that was impossible to do until only recently. In the early part of the twenty-first century, even questioning that narrative was sufficient for total ostracization and cancellation. Whereas now, claiming that Winston Churchill was the "chief villain of World War Two" or that a cabal of globalist Jews started the war against a group of sympathetic, ethnic Europeans is taken more seriously.[30]

Its symbolic apotheosis is revising the six million number put forth to express how many Jews were killed by Nazi Germany. The argument goes that Jews co-opted and manipulated the historical record to elevate their own suffering above other groups. Rather than claiming the "Holocaust never happened," revisionists now claim far less Jews were killed in the concentration camps than previously stated. However, whether the number is 271,000 or six million is irrelevant. The genocidal regime of Nazi Germany founded on racial theories of eugenics was irredeemable and rightfully destroyed.[31]

A similar idea has spread to practically every other realm of history and geopolitics. Yours truly can remember when the fashionable and faux-edgy interpretation of 9/11 was to say "Bush did it for oil" or something to that effect (pressing its adherents was always an exercise in obfuscation). Theories about 9/11 range from the towers being holograms to controlled demolition to a CIA-launched plot. In reality, they differ in degree based solely on how severe the theorist is suffering from schizophrenia. If it was controlled demolition, then what role did the planes play? Did the conspirators detonate the explosions as the planes collided or after? Were there not easier and less elaborate ways to get America involved in the Middle East (as the United States already was prior)?[32]

Now the trendy interpretation has morphed from "Bush did 9/11" to "Israel did 9/11" with all machinations of world politics credited to "the Jews." This lazy explanation only functions to further cede control to centralized authority. The responsibility and agency of individuals being removed as "What could I possibly do? The Jews control everything."[33]

The effect of these reevaluations will be enormous. The entire structure of the modern world order was founded on conclusions about who occupied "good and evil" during the early twentieth century. Upending these conclusions will create conditions for a new, likely more corrupted, future.

For the record, the result of the Second World War was enormously positive. Thankfully, racist ideas like eugenics were expunged with the defeat of Nazi Germany and Imperial Japan. It is fundamentally wrong to base societies on race. In every instance, the result is genocide and mass suffering. The Allies were justified in prosecuting and ultimately winning the Second World War.

While a healthy skepticism toward "expertise" or conventional wisdom is imminently prudent, as it was during the COVID-19 pandemic, it should not mutate into questioning every facet of the human experience. Doing so is a descent into mindless refutations of truth itself. Undermining every belief about the past, no matter how accurate, creates conditions for either the repetition of past atrocities or even more debased ones.

***

History is inquiry, and the results of this inquiry are ultimately shallow. All the fruits of humanity's genius—its gift for beauty, its power to play God and resurrect life from death—amount only to a tragic descent into senseless absurdity. The same species responsible for splitting the atom resigned to wearing useless masks on their face and erecting plexiglass barriers to stem aerosolized pathogens. In the decades and even centuries to come, better histories will be published about the COVID-19 pandemic. A history written while it is occurring will differ from one written five years later, or twenty-five years later or even one hundred years later. That being said, it is important to set the record straight now before it can be too distorted and leveraged by propagandists.

Additionally, while it may appear so superficially, none of what has been espoused in this book should be considered controversial. Affirmative action is racist. Biological men should not be allowed to compete athletically against women. Locking down the population over a cold virus was stupid.

Speaking the truth has never been fashionable, or easy or "cool." Those who proclaim it are often shouted down and demeaned. Being honest is uncomfortable and it is far easier to sit back and accept

155

conditions as they are—no matter how disastrous the consequences of staying silent may be. The stakes are no less than the maintenance of human liberty and survival of the species itself.

In fact, humans evolved to tell the truth. Ascending to the next evolutionary plateau depended on it. When the den of lions migrated closer and closer to the cave, the primate who communicated that the tribe need move to a different cave passed on their genes. Later on, hunter-gatherers who challenged their leadership for behaving idiotically and destructively, survived. Today, evolutionary progress depends on truth-tellers stating the seemingly obvious that others are either too scared or cognitively challenged to see. Funding experimentation to make deadly viruses more contagious is stupid and should stop. Destroying society and economic activity in response to a cold virus was stupid and should not be repeated.

Ultimately, the failure to speak the truth is rooted in fear. The tyrannical lockdowners acted out of fear for losing their relevancy. They acted out of fear of losing any iota of influence they previously had. In the case of the pandemic, despite the warnings, the begging, the entreatments to stop—to end the madness—they pressed on.

The origin of this study was a simple question. Why did people—regular individuals and leaders—act the way they did? Ultimately, its results are hollow in answering that question. We are no closer to their answers just as we are no closer to understanding why millions of Germans supported Hitler, or millions of Chinese idolized Mao Zedong. Indeed, there are episodes of madness and insanity that defy rational explanation. Like Hannah Arendt said, an examination of evil reveals nothing—as evil does not exist.[34]

As this book has accomplished, we can elucidate the reasons for why groups covered up their responsibility in perpetrating heinous disasters. We can illuminate the reasons why some thought it prudent to render viruses more transmissible to humans. We can even discern the motivations behind why a viral pandemic morphed into a nationwide racial riot. Yet, the reasons why family members were forbidden from visiting dying loved ones, or why some wished more children died from the disease, or why individuals were confined inside their homes to combat an aerosolized common cold is beyond this book's scope. As said before, maybe brighter minds than yours truly will someday uncover the reasons for this.

However, this study has revealed something—possibly even more profound. How do these episodes of madness end? How do we return life to peace and normalcy? How do we throw off the yoke of authoritarian despotism and reclaim our freedom? On these questions, the results of this inquiry are crystal clear. *Tyranny ends with the refusal of the subjugated.*

The COVID lockdowns did not end because leaders voluntarily lifted them. They did not end because the virus had been mitigated or eradicated. They did not end because leaders wished to set us free. They ended because of a collective refusal to abide by their mandates. They ended because regular people chose to violate their demands despite its consequences. They ended because we said so.

Finally, whether this Republic lives or dies, one thing remains true— its ideals. Its foundation to honor the individual and protect its freedom to pursue their own life remains unflinchingly true and just. The record will prove that the overall impact of the United States was positive. That it was a worthy cause—one worth dying for. That the thousands who perished at Omaha Beach, Salerno, Belleau Wood, and everywhere else soldiers spilt blood for the ideals of freedom was worth it. A fact no fatalist European, self-hating American or East Asian authoritarian can refute, despite their jealous scoffing or unintelligible ramblings.

In 1789, for the first time in human history, individuals granted powers and abilities to their government—rather than the other way around. For millennia, apart from brief respites in Ancient Greece, Rome, and thirteenth-century England, the *status quo* of human relations was nothing less than top-down authoritarianism. Rulers conferred to their people what they could and could not do.

The United States Constitution was not a document outlining what powers were to be vested in authority. Nor was it a dictation from above circumscribing the freedom of its people. Rather, the *people themselves* stipulated to government the scope of its power. In other words, we the people own and control our government. We the people decide when it has violated its permission. May we never forget this. May we never allow it to subjugate us again.

NOTES

MADNESS

1. Hannah Arendt, *Eichmann in Jerusalem: A Report on the Banality of Evil* (New York: Viking Press, 1963).

2. United States Global Leadership Coalition, "How COVID-19 Is Impacting Economies of Developing Countries," *USGLC*, accessed April 28, 2025; Caitlin O'Kane, "Texas Man Sentenced to 15 Months in Prison for Spreading COVID-19 Hoax on Facebook," *CBS News*, October 6, 2021.

3. John P. A. Ioannidis, "Infection Fatality Rate of COVID-19 Inferred from Seroprevalence Data," *Bulletin of the World Health Organization* 99, no. 1 (January 1, 2021): 19–33F.

4. Trust for America's Health, "U.S. Experienced Highest Ever Combined Rates of Deaths Due to Alcohol, Drugs, and Suicide During the COVID-19 Pandemic," *Trust for America's Health*, May 24, 2022; Feeding America, "60 Million People Turned to Charitable Food Assistance in 2020 amid COVID-19 Pandemic," *Feeding America*, September 1, 2021; Gene Falk et al., *Unemployment Rates During the COVID-19 Pandemic*, CRS Report R46554 (Washington, DC: Congressional Research Service, August 20, 2021).

5. Florian Ege, Giovanni Mellace, and Seetha Menon, "The Unseen Toll: Excess Mortality During COVID-19 Lockdowns," *Scientific Reports* 13 (October 31, 2023): Article 18745; Hashir Ali Awan et al., "Internet and Pornography Use During the COVID-19 Pandemic: Presumed Impact and What Can Be Done," *Frontiers in Psychiatry* 12 (April 2021): Article 623508.

6. Patrick Henry, "Economic Inequality Has Deepened during the Pandemic. That Doesn't Mean It Can't Be Fixed," *World Economic Forum*, April 7, 2022; Nick Timiraos, "March 2020: How the Fed Averted Economic Disaster," *The Wall Street Journal*, February 18, 2022.

7. Mark Thornton, "Cantillon Effects: Why Inflation Helps Some and Hurts Others," *Mises Wire*, March 11, 2022.

8. Meghan Roos, "Governor Whitmer Lifts Michigan's Safer at Home Order After Weeks of Protests, Says Restaurants Can Reopen Soon," *Newsweek,* June 1, 2020.

9. László Christián, Ákos Erdős, and Gergő Háló, "The Background and Repercussions of the George Floyd Case," *Cogent Social Sciences* 8, no. 1 (2022): Article 2082094.

10. Julia Musto, "Nearly 15M Deaths Worldwide Associated with COVID-19: WHO," *Fox News,* May 5, 2022; Joel Achenbach, "Covid was fourth leading cause of death in 2022, CDC data shows," *The Washington Post,* May 4, 2023.

11. Stephanie Armour, Brianna Abbott, Thomas M. Burton, and Betsy McKay, "What Derailed America's Covid Testing: Three Lost Weeks," *Wall Street Journal,* August 18, 2020.

12. Ioannidis, "Infection Fatality Rate of COVID-19."

13. Centers for Disease Control and Prevention, "Influenza (Flu)," *National Center for Health Statistics,* last reviewed February 12, 2024.

14. Centers for Disease Control and Prevention (CDC), "Excess Deaths Associated with COVID-19," *National Center for Health Statistics,* last reviewed March 28, 2024.

15. Gregory French et al., "Impact of Hospital Strain on Excess Deaths during the COVID-19 Pandemic—United States, July 2020–July 2021," *American Journal of Transplantation* 22, no. 2 (February 2022): 654–57; National Institute of Mental Health, "Youth Suicide Rates Increased During the COVID-19 Pandemic," *Science Updates,* last modified October 3, 2023.

16. Wilson Msemburi, Ariel Karlinsky, Victoria Knutson, et al., "The WHO Estimates of Excess Mortality Associated with the COVID-19 Pandemic," *Nature* 613 (2023): 130–137; David Leonhardt, "Four Years Later, the Covid Pandemic's Legacy," *New York Times,* March 11, 2024; Centers for Disease Control and Prevention, "Excess Deaths Associated with COVID-19," *National Center for Health Statistics.*

17. Edouard Mathieu et al., "Excess Mortality during the Coronavirus Pandemic (COVID-19)," *Our World in Data,* 2020.

18. Shiqi Tang et al., "Aerosol Transmission of SARS-CoV-2? Evidence, Prevention and Control," *Environment International* 144

(November 2020): 106039; Dyani Lewis, "Why the WHO Took Two Years to Say COVID Is Airborne," *Nature*, April 6, 2022.

19. S. Ding et al., "Aerosols from Speaking Can Linger in the Air for up to Nine Hours," *Building and Environment* 205 (November 2021): 108239; "Aerosol and Surface Stability of SARS-CoV-2 as Compared with SARS-CoV-1," *New England Journal of Medicine* 382, no. 16 (March 17, 2020): 1564–1567.

20. "Coronavirus (COVID-19) Vaccinations," *Our World in Data*, last modified April 30, 2025,

21. Yogesh N. Suryawanshi and Dhananjay A. Biswas, "Herd Immunity to Fight Against COVID-19: A Narrative Review," *Cureus* 15, no. 1 (January 9, 2023): e33575; Martin Kulldorff, Sunetra Gupta, and Jay Bhattacharya, *The Great Barrington Declaration*, October 4, 2020.

22. C. Murphy et al., "Effectiveness of Social Distancing Measures and Lockdowns for Reducing Transmission of COVID-19 in Non-Healthcare, Community-Based Settings," *Philosophical Transactions of the Royal Society A: Mathematical, Physical and Engineering Sciences* 381, no. 2257 (October 9, 2023): 20230132.

23. Timothy Puko and Dino Grandoni, "Massive Climate Change Protests in New York Aim to Turn Up Heat on Biden," *Washington Post*, updated September 18, 2023; Shannon Osaka, "What the World Would Look like without Fossil Fuels: A Thought Experiment Shows the Complexities of Phasing Out Oil, Gas and Coal," *Washington Post*, September 30, 2023.

24. "Guideline for Isolation Precautions: Preventing Transmission of Infectious Agents in Healthcare Settings," *Centers for Disease Control and Prevention*, accessed May 4, 2025.

25. Tedros Adhanom Ghebreyesus, "WHO Director-General's Opening Remarks at the Media Briefing on COVID-19 — 11 March 2020," *World Health Organization*, March 11, 2020; Daniel Dale, "Fact-Checking Trump's Repeated False Claim That Coronavirus Case Rise Is Due to Testing," *CNN*, July 9, 2020.

26. B. Wang et al., "Asymptomatic SARS-CoV-2 Infection by Age: A Global Systematic Review and Meta-analysis," *Pediatric Infectious Disease Journal* 42, no. 3 (March 1, 2023): 232–239; Sanya Mansoor,

"'We Opened Up Too Soon.' Atlanta Mayor Keisha Lance Bottoms Wants Georgia to Roll Back Reopening," *Time*, August 13, 2020.

27. Rick Rojas, "Trump Criticizes Georgia Governor for Decision to Reopen State," *New York Times*, April 22, 2020; Isaac Stanley-Becker, "Gov. Brian Kemp Sets Georgia on Aggressive Course to Reopen, Putting His State at Center of Deepening National Debate," *Washington Post*, April 21, 2020.

28. Reed Abelson, "Hospitals Are Reeling Under a New COVID Surge," *New York Times*, November 27, 2020; Sharon Otterman, "Covid Cases Surge in Upstate New York, Fueling Nationwide Rise," *New York Times*, December 3, 2021.

29. Amanda Macias, "Nearly 90% of the US Navy Hospital Ship in New York Is Empty amid Coronavirus Fight," *CNBC*, April 17, 2020; Guillaume Meille et al., "COVID-19 Admission Rates and Changes in US Hospital Inpatient and Intensive Care Unit Occupancy," *JAMA Health Forum* 4, no. 12 (December 1, 2023): e234206.

30. Friedrich Nietzsche, *Beyond Good and Evil: Prelude to a Philosophy of the Future*, trans. Walter Kaufmann (New York: Vintage Books, 1989), 89.

31. Joe Silverstein, "Podcaster Sam Harris Says COVID 'Could Have Killed All the Children' and He'd Still Have 'No F---ing Patience' for Vaccine Skeptics," *Fox News*, January 17, 2023.

32. Rebecca F. Wilson et al., "Unintentional Firearm Injury Deaths Among Children and Adolescents Aged 0–17 Years — National Violent Death Reporting System, United States, 2003–2021," *MMWR Morbidity and Mortality Weekly Report* 72 (2023): 1338–45; Canadian Paediatric Society, "Swimming and Water Safety for Young Children," *Paediatrics & Child Health* 8, no. 2 (February 2003): 117–20.

33. Jovana Pejovic, Cátia Severino, Marina Vigário, and Sónia Frota, "Prolonged COVID-19 Related Effects on Early Language Development: A Longitudinal Study," *Early Human Development* 181 (March 2023): 105205; David W. McCormick et al., "Deaths in Children and Adolescents Associated with COVID-19 and MIS-C in the United States," *Pediatrics* 148, no. 5 (November 2021): e2021052273.

34. Mark Duggan, Audrey Guo, and Andrew C. Johnston, "Unemployment during the Pandemic: How to Avoid Going for Broke,"

Stanford Institute for Economic Policy Research (SIEPR), September 2020; Shannon Palus, "Stop Yelling at Runners for Not Wearing Masks! It's Scientifically Ludicrous to Insist on This," *Slate*, April 30, 2020.

35. Dan Diamond, "A Mural of Tony Fauci Was Meant to Inspire Staff. Then NIH Took It Down," *Washington Post*, March 15, 2025.

36. Ed Yong, "Second Mutant Flu Paper Published," *Nature*, June 21, 2012; University of North Carolina Gillings School of Global Public Health, "Ralph S. Baric, PhD," accessed April 28, 2025.

37. Jon Cohen, "Scientists Strongly Condemn Rumors and Conspiracy Theories about Origin of Coronavirus," *Science*, February 19, 2020; Alexandra Stevenson, "Senator Tom Cotton Repeats Fringe Theory of Coronavirus Origins," *New York Times*, February 17, 2020; University of California San Diego Health, "Novel Coronavirus Circulated Undetected Months before First COVID-19 Cases in Wuhan, China," news release, March 18, 2021.

38. Erin Hale and Andy Penafuerte, "The Protests That Exposed Cracks in China's Middle-Class Dream," *Al Jazeera*, December 22, 2022.

39. Jonathan Chait, "Lying About the Lab-Leak Hypothesis Was Unjustified and Counterproductive," *Intelligencer*, February 28, 2023; Peter Knight, "COVID-19: Why Lab-Leak Theory Is Back Despite Little New Evidence," *The Conversation*, June 21, 2021.

40. Claire Shaffer, "Jon Stewart Says Lab-Leak Theory 'Is Not a Conspiracy' in Colbert Appearance," *Rolling Stone*, June 15, 2021.

41. Quint Forgey, "Fauci Endorses National Stay-at-Home Order: 'I Just Don't Understand Why We're Not Doing That,'" *Politico*, April 3, 2020.

42. Fred Guterl, "Dr. Fauci Backed Controversial Wuhan Lab with Millions of
U.S. Dollars for Risky Coronavirus Research," *Newsweek*, April 28, 2020; Jocelyn Kaiser, "House Panel Concludes COVID-19 Pandemic Came from Lab Leak," *Science*, December 3, 2024; Glenn Kessler, "Unpacking the Story about Fauci and Painful Experiments Involving Dogs," *Washington Post*, June 7, 2024.

43. The Editorial Board, "Fauci and Collins Trashed the COVID Lockdown Critics," *Wall Street Journal*, December 21, 2021; Josh Christenson, "NIH Director Admits Taxpayers Funded Gain-of-Function

Research in Wuhan Four Years after COVID Pandemic Began," *New York Post*, May 16, 2024; Jeff Zymeri, "Fauci Testimony on Gain-of-Function Research Was Inconsistent with Existing Intel, Says Ex-Director of National Intelligence," *National Review*, April 19, 2023.

44. Michael Specter, "How Anthony Fauci Became America's Doctor," *The New Yorker*, April 10, 2020.

45. Sheera Frenkel, Ben Decker and Davey Alba, "Virus Conspiracies Gain Steam as YouTube and Facebook Remove 'Plandemic,'" *New York Times*, May 20, 2020.

46. Nick R. Martin, "Infowars Host Owen Shroyer Debates 'Jewish Conspiracies' with Antisemitic YouTuber," *Hatewatch (Southern Poverty Law Center),* October 2, 2018.

47. Stephan Rosiny, "The Arab Spring: A Starting Point for Democratization?" *GIGA German Institute of Global and Area Studies,* 2012.

48. Edmund W. J. Lee, Huanyu Bao, Yixi Wang, and Yi Torng Lim, "From Pandemic to Plandemic: Examining the Amplification and Attenuation of COVID-19 Misinformation on Social Media," *Telematics and Informatics* 328 (July 2023): 115979.

49. Madelon van Gerwen et al., "Risk Factors and Outcomes of COVID-19 in New York City: A Retrospective Cohort Study," *Journal of Medical Virology* 93, no. 2 (February 2021): 907–15; Centers for Disease Control and Prevention, "COVID-19: People with Certain Risk Factors," last reviewed February 29, 2024.

50. Laura Joszt and Chris Mazzolini, "Johns Hopkins Working Paper Says COVID-19 Lockdowns Not Worth It, Sparks Fierce Debate," *Medical Economics*, February 7, 2022.

51. Pew Research Center, "Public Trust in Scientists and Views on Their Role in Policymaking," November 14, 2024.

52. Michaeleen Doucleff, "Why It's Difficult for Viruses to Turn into Deadly Pandemics," *NPR*, May 29, 2018.

53. Betzaida Tejada-Vera and Ellen A. Kramarow, *COVID-19 Mortality in Adults Aged 65 and Over: United States, 2020*, NCHS Data Brief No. 446 (Hyattsville, MD: National Center for Health Statistics, October 2022).

54. American College of Veterinary Pathologists, "Bird Flu Fact Sheet," accessed April 28, 2025; Lauren Pelley, "Scientists Warn H5N1 Bird Flu Needs Urgent Action to Prevent a Human Pandemic," *CBC News*, May 4, 2024.

55. Nicholas Bariyo, "Mystery Disease Linked to Bats Kills Scores in Congo,"

Wall Street Journal, February 27, 2025.

56. Pew Research Center, "Americans' Trust in Scientists, Other Groups Declines," February 15, 2022.

57. Brian Kennedy and Alec Tyson, "Americans' Trust in Scientists, Positive Views of Science Continue to Decline," *Pew Research Center*, November 14, 2023; Quentin Fottrell, "Defiant Fauci Tells InStyle: 'I Don't Regret Anything I Said' and Talks about Being Persona Non Grata in Trump's White House," *MarketWatch*, July 20, 2020.

58. Thomas Robert Malthus, *An Essay on the Principle of Population* (London: J. Johnson, 1798); Paul R. Ehrlich, *The Population Bomb* (New York: Ballantine Books, 1968).

59. Mark Landler and Stephen Castle, "Behind the Virus Report That Jarred the U.S. and the U.K. to Action," *New York Times*, March 17, 2020; Nature, "COVID's 'Patient Zero': Why It's So Hard to Find the First Case," *Nature*, January 21, 2020; Neil M. Ferguson et al., *Impact of Non-Pharmaceutical Interventions (NPIs) to Reduce COVID-19 Mortality and Healthcare Demand*, Imperial College COVID-19 Response Team, March 16, 2020; Peter St. Onge, with the collaboration of Gaël Campan, *When the Cure Is Worse than the Disease: The Burden of Lockdowns in Canada*, Montreal Economic Institute, June 2020.

60. Toby Young, "Why Can't We Talk about the Great Barrington Declaration?" *The Spectator*, October 17, 2020.

61. Ted Mann, "U.S. Coronavirus Cases Rise by Nearly 40,000," *Wall Street Journal*, September 23, 2020; Will Stone, "U.S. Coronavirus Cases Surpass Summer Peak and Are Climbing Higher Fast," *NPR*, October 27, 2020; Natural News Editors, "Tucker Carlson Triggers CNN by Pointing Out COVID-19 Death Count Inflation," *Natural News*, January 28, 2021.

62. Centers for Disease Control and Prevention, "2023–2024 Influenza Season Summary: Influenza Severity Assessment, Burden and

Burden Prevented," last modified November 22, 2024; Ioannidis, "Infection Fatality Rate of COVID-19."

63. Lucía L. Maldonado, Adriana M. Bertelli, and Lucas Kamenetzky, "Molecular Features Similarities Between SARS-CoV-2, SARS, MERS and Key Human Genes Could Favour the Viral Infections and Trigger Collateral Effects," *Scientific Reports* 11 (2021): 4108.

64. Loh, Christine. *At the Epicentre: Hong Kong and the SARS Outbreak*. Vol.

1. Hong Kong University Press, 2004; J. D. Cherry and P. Krogstad, "SARS: The First Pandemic of the 21st Century," *Pediatric Research* 56, no. 1 (July 2004): 1–5.

65. J. Zhai et al., "Real-Time Polymerase Chain Reaction for Detecting SARS Coronavirus, Beijing, 2003," *Emerging Infectious Diseases* 10, no. 2 (February 2004): 300–303.

66. Nicole Jawerth, "How Is the COVID-19 Virus Detected Using Real Time RT–PCR?" *IAEA Bulletin* 61, no. 2 (June 2020); Jordan Green, "US Government No Longer Offering Free COVID-19 Tests: Where to Get an At-Home Test," *Memphis Commercial Appeal*, March 19, 2025.

67. World Health Organization, "WHO Coronavirus (COVID-19) Dashboard," accessed April 28, 2025; Centers for Disease Control and Prevention, "Basic Information About SARS," archived page, last reviewed February 10, 2004.

68. Alex Pattakos, "The Paradox of 'Trump Derangement Syndrome': Finding Meaning in the Space between Stimulus and Response," *Psychology Today*, September 5, 2024.

69. USAFacts, "2019 Economy in Review: GDP, Employment, Income, and Trade," accessed April 28, 2025.

70. RealClearPolitics, "Trump vs. Biden: Top Battleground States (2020 vs. 2016)," accessed April 28, 2025.

71. Carol Leonnig and Philip Rucker, *I Alone Can Fix It: Donald J. Trump's Catastrophic Final Year* (New York: Penguin Press, 2021).

72. Sharon Begley, "With Ventilators Running Out, Doctors Say the Machines Are Overused for Covid-19," *STAT*, April 8, 2020.

73. Maggie Haberman, "Trump Has Given Unusual Leeway to Fauci, but Aides Say He's Losing His Patience," *New York Times*, March 23, 2020.

74. Rick Rojas and Richard Fausset, "'I Am Beyond Disturbed':
Internal Dissent as States Reopen Despite Virus," *New York Times*, April
21, 2020.

75. James Madison, *Federalist No. 45*, in *The Federalist Papers*, ed.
Clinton Rossiter (New York: Signet Classics, 2003), 292–297.

76. Robert Tait, "PEN America Calls Out US Politicians for
'Sustained' Censorship of Writers," *The Guardian*, April 24, 2025; U.S.
Department of Justice, "Addressing Police Misconduct: Laws Enforced
by the Department of Justice," accessed April 28, 2025.

77. Peter Baker, "Trump Vows to 'Fix It,' Whether It's Still Broken
or Not," *New York Times*, November 4, 2024.

78. Karl W. Smith, "Trump's Economy Really Was Better Than
Obama's," *Bloomberg*, October 30, 2020; Chuck Jones, "Obama's 2009
Recovery Act Kicked Off Over 10 Years of Economic Growth," *Forbes*,
February 17, 2020.

79. Milton Friedman and Rose D. Friedman, *Free to Choose: A
Personal Statement* (New York: Harcourt Brace Jovanovich, 1980).

80. Frank Muci, "Why Did Venezuela's Economy Collapse?"
Economics Observatory, September 23, 2024; Central Intelligence
Agency, "The Republics of the Former USSR: The Outlook for the Next
Year," September
1991, table 5, Cold War International History Project, *Documents
and Papers.*

81. Centers for Disease Control and Prevention, "Final Data on
COVID-19 Mortality," *National Center for Health Statistics*, accessed
April 28, 2025.

82. Caitlin Oprysko, "Trump Says He Told Kemp: 'I Totally
Disagree' with Move to Reopen Georgia," *Politico*, April 22, 2020.

83. John Ferling, *Adams vs. Jefferson: The Tumultuous Election of
1800* (New York: Oxford University Press, 2004).

84. The Pulitzer Prizes, "Staffs of The New York Times and The
Washington Post," accessed April 28, 2025.

85. National Archives, "President Dwight D. Eisenhower's Farewell
Address," accessed April 28, 2025.

86. Jason DeRose, "Religion Is Less Important in the Lives of
Americans, Report Finds," *NPR*, May 16, 2023.

87. National Science Foundation, "Mind-Body Connection Is Built into the Brain, Study Suggests," accessed April 28, 2025; Alastair Wilson, "What Existed Before the Big Bang?" *BBC Future*, January 5, 2022.

88. Faisal Islam, "We'll Need Universal Basic Income – AI 'Godfather'," *BBC News*, May 18, 2024; Kevin Okemwa, "Bill Gates Says, 'We Weren't Born to Do Jobs. AI Will Replace Humans for Most Things,'" *Windows Central*, April 4, 2025.

89. Herodotus, *The Histories*, trans. Robin Waterfield (Oxford: Oxford University Press, 1998).

90. Sean McMeekin, *Stalin's War: A New History of World War II* (New York: Basic Books, 2021); Chris Bellamy, *Absolute War: Soviet Russia in the Second World War* (New York: Alfred A. Knopf, 2007).

91. George C. Herring, *Aid to Russia, 1941–1946: Strategy, Diplomacy, the Origins of the Cold War* (New York: Columbia University Press, 1973); Craig L. Symonds, *Neptune: The Allied Invasion of Europe and the D-Day Landings* (New York: Oxford University Press, 2014).

92. Michael Lewis, *The Premonition: A Pandemic Story* (New York: W. W. Norton & Company, 2021).

93. Ram Dass, *Be Here Now* (San Cristobal, NM: Lama Foundation, 1971); Eckhart Tolle, *The Power of Now: A Guide to Spiritual Enlightenment* (Novato, CA: New World Library, 1999).

94. Henry, "Economic Inequality Has Deepened during the Pandemic."; University of Nebraska Medical Center, "How the Pandemic Messed with Our Perception of Time," August 8, 2023; Navigate Affordable Housing Partners, "President's Coronavirus Guidelines: 15 Days to Slow the Spread," accessed April 28, 2025.

95. Aleksandr I. Solzhenitsyn, *The Gulag Archipelago, 1918–1956: An Experiment in Literary Investigation*, trans. Thomas P. Whitney and Harry Willetts (New York: Harper & Row, 1974–1978).

96. François Ponchaud, *Cambodia: Year Zero*, trans. Nancy Amphoux (New York: Holt, Rinehart and Winston, 1978).

97. Dith Pran, comp., *Children of Cambodia's Killing Fields: Memoirs by Survivors*, ed. Kim DePaul, intro. Ben Kiernan (New Haven: Yale University Press, 1997).

98. Allan B. de Guzman et al., "Who Says Aging Is Lonely? A Phenomenology of Filipino Older Adults' Experiences of Happiness

When Joining International Group Tours," *Educational Gerontology* 45, no. 6 (2019): 365–76.

99. C. G. Jung, *Psychology and Alchemy*, trans. R. F. C. Hull (Princeton, NJ: Princeton University Press, 1968), 42.

100. Thucydides, *History of the Peloponnesian War*, trans. Richard Crawley (New York: Modern Library, 1951), 1.22.

THE ROAD TO WUHAN

1. Henry Kissinger, *On China* (New York: Penguin Press, 2011).

2. W. Travis Hanes III and Frank Sanello, *Opium Wars: The Addiction of One Empire and the Corruption of Another* (Naperville, IL: Sourcebooks, 2002).

3. Norman Ohler, *Blitzed: Drugs in the Third Reich*, trans. Shaun Whiteside (New York: Houghton Mifflin Harcourt, 2017); Mark Pendergrast, *Uncommon Grounds: The History of Coffee and How It Transformed Our World* (New York: Basic Books, 2019); UC Berkeley Center for the Science of Psychedelics, *"Acid Dreams: The Complete Social History of LSD, the CIA, the Sixties, and Beyond,"* accessed April 28, 2025.

4. Julia Lovell, *The Opium War: Drugs, Dreams, and the Making of China* (New York: Overlook Press, 2014); Sebastian Rotella, "Fentanyl Pipeline: How a Chinese Prison Helped Fuel a Deadly Drug Crisis in the United States," *ProPublica*, April 23, 2025.

5. William T. Rowe. *Hankow: Commerce and society in a Chinese city, 1796-1889.* (Palo Alto: Stanford University Press, 1992).

6. Lewis Pyenson and Susan Sheets-Pyenson, *Servants of Nature: A History of Scientific Institutions, Enterprises and Sensibilities* (New York: W.W. Norton, 1999).

7. Albert Einstein. "On the electrodynamics of moving bodies." *Annalen der physik* 17, no. 10 (1905): 891-921.

8. Emily Garvanovic, "Miasmatic Theory," *ESSAI* 12 (2014): Article 18.

9. John M. Barry, *The Great Influenza: The Story of the Deadliest Pandemic in History* (New York: Viking, 2004).

10. Katie Simmons, "China's Government May Be Communist, but Its People Embrace Capitalism," *Pew Research Center*, October 10, 2014.

11. Andrew Szamosszegi and Cole Kyle, *An Analysis of State-Owned Enterprises and State Capitalism in China* (Washington, DC: U.S.–China Economic and Security Review Commission, 2011); Tom Doctoroff, *What Chinese Want: Culture, Communism, and China's Modern Consumer* (New York: Palgrave Macmillan, 2012).

12. Lucian W. Pye, "Liberalization in China: Can Economics Be the Engine of Political Change," *Fletcher Forum of World Affairs* 12 (1988): 221.

13. Kissinger, *On China, 5-22.*

14. Xufeng Fang, *The Impact of Overseas Study Experiences on Chinese Students' Attitudes Toward the United States* (PhD diss., City University of New York, 2023); Gardner Bovingdon, *The Uyghurs: Strangers in Their Own Land* (New York: Columbia University Press, 2010), 3; Jane Ardley, *The Tibetan Independence Movement: Political, Religious and Gandhian Perspectives* (London: Routledge, 2003); Ravi Kanbur and Xiaobo Zhang, "Which Regional Inequality? The Evolution of Rural–Urban and Inland–Coastal Inequality in China from 1983 to 1995," *Journal of Comparative Economics* 27, no. 4 (1999): 686–701.

15. Kevin Carrico, *The Great Han: Race, Nationalism, and Tradition in China Today* (Oakland: University of California Press, 2017).

16. William T. Rowe, *China's Last Empire: The Great Qing* (Cambridge: Harvard University Press, 2009).

17. Jun Fu, "Big Contract in Small Village—Xiaogang Village and Rural Reforms," in *China's Pathways to Prosperity: Abductive Reflections on Reforms and Opening-Up* (Singapore: Springer Nature Singapore, 2025), 33–69; Frank Dikötter, *Mao's Great Famine: The History of China's Most Devastating Catastrophe, 1958–1962* (New York: Walker & Company, 2010); Yang Jisheng, *The World Turned Upside Down: A History of the Chinese Cultural Revolution*, trans. Stacy Mosher and Guo Jian (New York: Farrar, Straus and Giroux, 2021).

18. Richard Evans, *Deng Xiaoping and the Making of Modern China* (New York: Viking, 1994).

19. David O. Shullman, "How China Is Exploiting the Pandemic to Export Authoritarianism," *War on the Rocks*, March 31, 2020.

20. Talha Burki, "China's Successful Control of COVID-19," *The Lancet Infectious Diseases* 20, no. 11 (2020): 1240–41.

21. T. L. Xu, M. Y. Ao, X. Zhou, et al., "China's Practice to Prevent and Control COVID-19 in the Context of Large Population Movement," *Infectious Diseases of Poverty* 9 (2020): 115; VOA News, "What Is China's 'Zero-COVID' Policy?" *VOA News*, November 28, 2022; James Gregory, "China Deactivates National Covid Tracking App," *BBC News*, December 12, 2022.

22. Andrew Mark Miller, "China Arrests 9 Coronavirus Lockdown Enforcers after Video Shows Them Beating Civilians," *Fox News*, November 11, 2022; Junwei Yang and Timothy Reuter, "3 Ways China Is Using Drones to Fight Coronavirus," *World Economic Forum*, March 16, 2020; Linda Lew, "China's Vast 14,000-Bed Covid Isolation Center Revealed in Drone Footage," *Bloomberg*, September 21, 2022; Joe McDonald, "As Shanghai Lockdown Continues, Residents Face Food and Supply Shortages," *Associated Press*, April 7, 2022.

23. Kim Hjelmgaard, Eric J. Lyman, and Deirdre Shesgreen, "This Is What China Did to Beat Coronavirus. Experts Say America Couldn't Handle It," *USA Today*, April 1, 2020; Yasheng Huang, Meicen Sun, and Yuze Sui, "How Digital Contact Tracing Slowed Covid-19 in East Asia," *Harvard Business Review*, April 15, 2020.

24. World Health Organization, *Report of the WHO-China Joint Mission on Coronavirus Disease 2019 (COVID-19), 16–24 February 2020* (Geneva: WHO, 2020); Amy Qin, "China May Be Beating the Coronavirus, at a Painful Cost," *New York Times*, March 7, 2020, updated March 10, 2020; Myah Ward, "15 Times Trump Praised China as Coronavirus Was Spreading Across the Globe," *Politico*, April 15, 2020.

25. Robbie Griffiths, "China Has Stopped Publishing Daily COVID Data amid Reports of a Huge Spike in Cases," *NPR*, December 25, 2022; Associated Press, "China Reduces COVID-19 Case Number Reporting as Virus Surges," *NPR*, December 14, 2022.

26. Selam Gebrekidan, Matt Apuzzo, Amy Qin, and Javier C. Hernández, "In Hunt for Virus Source, W.H.O. Let China Take Charge," *New York Times*, November 2, 2020.

27. Hinnerk Feldwisch-Drentrup, "How WHO Became China's Coronavirus Accomplice," *Foreign Policy*, April 2, 2020.

28. Michael D. Bordo, *The Operation and Demise of the Bretton Woods System: 1958 to 1971*, NBER No. 23189 (Cambridge, MA: National Bureau of Economic Research, February 2017).

29. Linda Qiu, Bill Marsh, and Jon Huang, "The President vs. the Experts: How Trump Played Down the Coronavirus," *New York Times*, March 18, 2020.

30. World Health Organization, "WHO, China Leaders Discuss Next Steps in Battle Against Coronavirus Outbreak," news release, January 28, 2020; Owen Dyer, "Covid-19: China Pressured WHO Team to Dismiss Lab Leak Theory, Claims Chief Investigator," *BMJ* 374 (2021): n2023.

31. Dali L. Yang, *Wuhan: How the COVID-19 Outbreak in China Spiraled Out of Control* (New York: Oxford University Press, 2024), 170; Jessica Brandt, Bret Schafer, Elen Aghekyan, Valerie Wirtschafter, and Adya Danaditya, *Winning the Web: How Beijing Exploits Search Results to Shape Views of Xinjiang and COVID-19* (Washington, DC: Brookings Institution, May 2022).

32. Chia C. Wang et al., "Airborne Transmission of Respiratory Viruses," *Science* 373, no. 6558 (2021): eabd9149.

33. Paul Gordon Lauren, *Power and Prejudice: The Politics and Diplomacy of Racial Discrimination* (Boulder, CO: Westview Press, 1988).

34. Yang, *Wuhan*, 36-38.

35. Centers for Disease Control and Prevention, "Fact Sheet: Basic Information About SARS," May 3, 2005; World Health Organization, "SARS Outbreak Contained Worldwide," news release, July 5, 2003.

36. Mohamed A. Farrag et al., "SARS-CoV-2: An Overview of Virus Genetics, Transmission, and Immunopathogenesis," *International Journal of Environmental Research and Public Health* 18, no. 12 (June 10, 2021): 6312.

37. Yousef Alimohamadi, Maryam Taghdir, and Mojtaba Sepandi, "Estimate of the Basic Reproduction Number for COVID-19: A Systematic Review and Meta-analysis," *Journal of Preventive Medicine and Public Health* 53, no. 3 (May 2020): 151–157; Ying Liu et al., "The Reproductive Number of COVID-19 Is Higher Compared to SARS Coronavirus," *Journal of Travel Medicine* 27, no. 2 (March 13, 2020): taaa021.

38. The Newsroom, "Scientists Wait for Next Mass Killer to Spill Over from Nature," *The Yorkshire Post*, February 12, 2013; "Alert over Hong Kong 'Super-Flu,'" *BBC News*, March 13, 2003; Tom Huddleston Jr., "This Hotel Is Infamous as Ground Zero for a SARS 'Super Spreader' in the 2003 Outbreak—Here's What Happened," *CNBC*, February 16, 2020.

39. "Sars Illness 'Not Under Control,'" *BBC News*, April 11, 2003.

40. CBC Sports, "SARS: Women's World Cup Moved Out of China," *CBC*, May 3, 2003.

41. Jonathan Shaw, "The SARS Scare: A Cautionary Tale of Emerging Disease Caught in the Act," *Harvard Magazine*, March–April 2007; World Health Organization, "China's Latest SARS Outbreak Has Been Contained, but Biosafety Concerns Remain – Update 7," *Disease Outbreak News*, May 18, 2004.

42. Annelies Wilder-Smith et al., "Can We Contain the COVID-19 Outbreak with the Same Measures as for SARS?" *The Lancet Infectious Diseases* 20, no. 5 (2020): e102–e107.

43. Hawaii State Department of Health, "Coronaviruses (Common Cold Viruses)," Disease Outbreak Control Division, accessed April 30, 2025.

44. "About Respiratory Illnesses," *Centers for Disease Control and Prevention*, accessed May 4, 2025.

45. J. Zhai et al., "Real-Time Polymerase Chain Reaction for Detecting SARS Coronavirus, Beijing, 2003," *Emerging Infectious Diseases* 10, no. 2 (February 2004): 300–303.

46. S.L. Emery et al., "Real-Time Reverse Transcription–Polymerase Chain Reaction Assay for SARS-Associated Coronavirus," *Emerging Infectious Diseases* 10, no. 2 (2004): 311–316; S. Cheng, C. Fockler, W.M. Barnes, and R. Higuchi, "Effective Amplification of Long Targets from Cloned Inserts and Human Genomic DNA," *Proceedings of the National Academy of Sciences of the United States of America* 91, no. 12 (June 7, 1994): 5695–5699.

47. Subbu Dharmaraj, MS, "The Basics: RT-PCR," *Thermo Fisher Scientific*, accessed April 30, 2025.

48. F.S. Iliescu et al., "Point-of-Care Testing—The Key in the Battle against SARS-CoV-2 Pandemic," *Micromachines* 12, no. 12 (November 27, 2021): 1464.

49. Charles Krauthammer, "Bush Derangement Syndrome Is Spreading," *Deseret News*, December 7, 2003.

50. Steven Taylor, "The Psychology of Pandemics," *Annual Review of Clinical Psychology* 18 (2022): 581–609; Jacob Sullum, "Senseless Restrictions on Outdoor Activities Undermine the Goal of Curbing COVID-19," *Reason*, November 25, 2020.

51. John Paul Stevens, "Repeal the Second Amendment," *The New York Times*, March 27, 2018; Joe Silverstein, "Podcaster Sam Harris: If COVID Killed More Children There'd Be 'No F---ing Patience' for Vaccine Skeptics," *Fox News*, January 17, 2023.

52. Danson Cheong, "China Signals No Change to Zero-Covid-19 Policy Amid Mass Protests That Challenge Xi," *The Straits Times*, November 26, 2024; Chen, Jing. *Useful Complaints: How Petitions Assist Decentralized Authoritarianism in China.* (New York: Rowman & Littlefield, 2016).

53. Stein Ringen, *The Perfect Dictatorship: China in the 21st Century* (Hong Kong: Hong Kong University Press, 2016).

54. John Pomfret, "Outbreak Gave China's Hu an Opening: President Responded to Pressure Inside and Outside Country on SARS," *The Washington Post*, May 12, 2003.

55. Dave Gershgorn, "China's 'Sharp Eyes' Program Aims to Surveil 100% of Public Space," *One Zero*, March 2, 2021; John Pomfret, "China Feels Side Effects From SARS: Political Fallout Follows Coverup," *The Washington Post*, May 1, 2003.

56. "SARS Moves China Toward Transparency," *The Washington Times*, July 17, 2003.

57. Yanzhong Huang, "The SARS Epidemic and Its Aftermath in China: A Political Perspective," in *Learning from SARS: Preparing for the Next Disease Outbreak: Workshop Summary*, ed. S. Knobler, A. Mahmoud, S. Lemon, et al. (Washington, DC: National Academies Press, 2004).

58. Huang, "The SARS Epidemic and Its Aftermath in China: A Political Perspective"; Yanzhong Huang. "Pursuing Health as Foreign Policy: The

Case of China." *Indiana Journal of Global Legal Studies* 17, no. 1 (2010): 105–46.

59. Pomfret, "Outbreak Gave China's Hu an Opening"

60. Pomfret, "China Feels Side Effects From SARS"

61. Joe Havely, "Getting to Know Hu: Chinese City Does Little to Trumpet Connections to Its Most Famous Son," *Al Jazeera*, October 19, 2007.

62. Robert Lawrence Kuhn, *Jiang Zemin* (New York: Crown Publishers, 2004).

63. Kuhn, *Jiang Zemin,* 161.

64. "Tiananmen Square Protest Death Toll 'Was 10,000'," *BBC News,* December 23, 2017.

65. Kuhn, *Jiang Zemin,* 145-176.

66. Pomfret, "China Feels Side Effects From SARS"; Huang, "The SARS Epidemic and Its Aftermath in China: A Political Perspective"

67. Yang, *Wuhan,* 36-38.

68. M. A. Benitez, "Beijing Doctor Alleges SARS Cases Cover-Up in China," *The Lancet* 361, no. 9366 (April 19, 2003): 1357.

69. Amy Qin, "Jiang Yanyong, Who Helped Expose China's SARS Crisis, Dies at 91," *The New York Times,* March 14, 2023.

70. Huang, "The SARS Epidemic and Its Aftermath in China: A Political Perspective"

71. J.W. LeDuc and M.A. Barry, "SARS, the First Pandemic of the 21st Century," *Emerging Infectious Diseases* 10, no. 11 (November 2004): e26.

72. Shaw, "The SARS Scare"

73. S.E. Richardson, R. Tellier, and J. Mahony, "The Laboratory Diagnosis of Severe Acute Respiratory Syndrome: Emerging Laboratory Tests for an Emerging Pathogen," *Clinical Biochemistry Review* 25, no. 2 (May 2004): 133–141.

74. Shaw, "The SARS Scare"

75. World Health Organization, "*What We Do.*"

76. World Health Organization, "China's Latest SARS Outbreak Has Been Contained"

77. "China Coronavirus," *Worldometer*, accessed April 30, 2025.

78. Wuchun Cao, Liqun Fang, and Dan Xiao, "RETRACTED: What We Have Learnt from the SARS Epidemics in Mainland China?" *Global Health Journal* (2019); Kelly-Leigh Cooper, "China Coronavirus: The Lessons Learned from the SARS Outbreak," *BBC News*, January 23, 2020.

79. Shawn Yuan, "Inside the Early Days of China's Coronavirus Cover-Up," *The Big Story*, May 1, 2020; Cindy Carter, "On the Fifth Anniversary of COVID Whistleblower Dr. Li Wenliang's Death: 'The World Hasn't Gotten Any Better. If Anything, It's Even More Insane,'" *China Digital Times*, February 7, 2025; Dan De Luce, Robert Windrem, and Abigail Williams, "The Pandemic Shows WHO Lacks Authority to Force Governments to Divulge Information, Experts Say," *NBC News*, May 9, 2020.

80. P. Liu, Y. Guo, X. Qian, S. Tang, Z. Li, and L. Chen, "China's Distinctive Engagement in Global Health," *The Lancet* 384, no. 9945 (August 30, 2014): 793–804.

81. Robert F. Kennedy Jr., *The Wuhan Cover-Up: How the Chinese Government and Big Pharma Conspired to Hide the Origins of the COVID-19 Pandemic* (New York: Skyhorse Publishing, 2024), Kindle edition, 254-260; Victoria Bela, "China's Wuhan Virology Institute Creates Nasal Covid-19 Vaccine for 'Future Pandemics,'" *South China Morning Post*, September 11, 2024.

82. Charles Calisher et al., "Statement in Support of the Scientists, Public Health Professionals, and Medical Professionals of China Combatting COVID-19," *The Lancet* 395, no. 10226 (2020): e42–e43; Dan Diamond, "A Mural of Anthony Fauci Was Meant to Inspire Staff. Then NIH Took It Down," *The Washington Post*, March 15, 2025.

83. Stanley Fish, "Revisiting Affirmative Action, With Help From Kant," *The New York Times Opinionator*, January 14, 2007; J. Grossman, S. Tomkins,

L. Page, et al., "The Disparate Impacts of College Admissions Policies on Asian American Applicants," *Scientific Reports* 14 (2024): 4449; Alexandra Stevenson, "Senator Tom Cotton Repeats Fringe Theory of Coronavirus Origins," *The New York Times*, February 2020; James B. Meigs, "The Lab-Leak Theory Cover-Up," *Commentary Magazine*, July/August 2021.

84. "Classified State Department Documents Credibly Suggest COVID-19 Lab Leak, Wenstrup Pushes for Declassification," *Select Subcommittee on the Coronavirus Pandemic*, May 7, 2024.

85. Gordon G. Chang, "China Deliberately Spread the Coronavirus: What Are the Strategic Consequences?" *Hoover Institution*, December 9, 2020.

86. Lynette H. Ong, "China's Massive Protests Are the End of a Once-Trusted Governance Model," *Foreign Policy*, November 28, 2022; Vijdan Mohammad Kawoosa, Engen Tham, Joyce Zhou, and Yew Lun Tian, "How COVID Protests Spread across China: A Timeline of Recent Protests," *Reuters*, November 28, 2022, updated November 29, 2022; Karson Yiu, "How a Deadly Apartment Fire Fueled Anti-Zero-COVID Protests Across China: ANALYSIS," *The Hong Kong Free Press*, November 27, 2022.

87. Julie Zhu, Yew Lun Tian, and Engen Tham, "Insight: How China's New No. 2 Hastened the End of Xi's Zero-COVID Policy," *Reuters*, March 2, 2023.

88. Tong Tung Yeng. "Chinese Authorities Shut Down Exhibition for Artwork Referencing the Lockdown." *ArtAsiaPacific*, May 6, 2023.

89. Vivian Wang, "'Let's Not Talk About It': 5 Years Later, China's Covid Shadow Lingers," *The New York Times*, March 13, 2025.

90. Frederik Kelter, "China Tries to 'Bury the Memory' and Trauma of Zero-COVID Era," *Al Jazeera*, December 21, 2023.

91. Sun Tzu, *The Art of War*, trans. Lionel Giles (New York: Dover Publications, 2002), Initial Estimations.

THE GOOD DOCTOR

1. "News Summary; Friday, June 5, 1981," *The New York Times*, June 5, 1981, Section B, Page 1.

2. Anthony S. Fauci, *On Call: A Doctor's Journey in Public Service* (New York: Viking, 2024), 33-36.

3. Fauci, *On Call*, 3-29.

4. W.A. Haseltine, "Molecular Biology of the Human Immunodeficiency Virus Type 1," *The FASEB Journal* 5 (1991): 2349–2360.

5. Fauci, *On Call*, 37-39.

6. Daniel B. Hrdy, "Cultural Practices Contributing to the Transmission of Human Immunodeficiency Virus in Africa", *Reviews of Infectious Diseases*, Volume 9, Issue 6, November 1987, Pages 1109–1119.

7. Fauci, *On Call*, 47-48.

8. Fauci, *On Call*, 49-67.

9. Fauci, *On Call*, 75-18.

10. Larry Kramer, "An Open Letter to Dr. Anthony Fauci," *The Village Voice*, May 31, 1988; Jacob Bernstein, "At Larry Kramer's Memorial, a Gathering of Friends and Enemies," *The New York Times*, June 27, 2023, updated June 29, 2023.

11. Adrian R. Lewis, *Omaha Beach: A Flawed Victory* (Chapel Hill: University of North Carolina Press, 2001).

12. Amelia Goranson, Paschal Sheeran, Julia Katz, and Kurt Gray, "Doctors Are Seen as Godlike: Moral Typecasting in Medicine," *Social Science & Medicine* 258 (2020): 113008.

13. Rachel Hammer, "The God Complex," *Academic Medicine* 87, no. 6 (June 2012): 775.

14. Carlie Porterfield, "Dr. Fauci On GOP Criticism: 'Attacks On Me, Quite Frankly, Are Attacks On Science'," *Forbes*, June 9, 2021.

15. Nicoli Nattrass, "AIDS Denialism vs. Science," *Skeptical Inquirer* 31, no. 5 (2007): 31–35; Fauci, *On Call*, 153-157.

16. Françoise Barré-Sinoussi, "HIV as the Cause of AIDS," *The Lancet* 348, no. 9019 (1996): 31–35.

17. Monya Baker, "1,500 Scientists Lift the Lid on Reproducibility," *Nature* 533 (2016): 452–454.

18. John Cook, "It's True: 97% of Research Papers Say Climate Change Is Happening," *The Conversation*, May 15, 2013; Peter Shawn Taylor, "Make Skepticism Great Again: The Replication Crisis in Science and What It Means for the Rest of Us," *National Post*, December 2, 2021.

19. P. Chigwedere and M. Essex, "AIDS Denialism and Public Health Practice," *AIDS and Behavior* 14 (2010): 237–247.

20. Fauci, *On Call,* 153-157.

21. Fauci, *On Call,* 253-342.

22. Will Stone, "1st U.S. Case of Coronavirus Confirmed in Washington State," *NPR,* January 22, 2020.

23. Fauci, *On Call,* 345-347.

24. Kevin Liptak, "Trump Declares National Emergency – and Denies Responsibility for Coronavirus Testing Failures," *CNN,* March 13, 2020; Daniel Lippman, "The Purell Presidency: Trump Aides Learn the President's Real Red Line," *Politico,* July 7, 2019.

25. Colm Quinn, "Trump Extends Coronavirus Lockdown Until April 30," *Foreign Policy,* March 30, 2020.

26. Eileen Drage O'Reilly, "Americans Still Have Low Risk of Coronavirus Infection, CDC Says," *Science,* January 27, 2020; Kyle Smith, "Anthony Fauci's Misadventures in Fortune Telling," *National Review,* April 20, 2021.

27. Steve Marsh, "Mike Osterholm's Last Stand," *Minnesota Monthly,* July 19, 2020; Frank E. Lockwood, "Cotton's Theory on Virus Origin No Longer Dismissed," *Arkansas Democrat-Gazette,* May 30, 2021.

28. Fred Guterl, "Dr. Fauci Backed Controversial Wuhan Lab with U.S. Dollars for Risky Coronavirus Research," *Newsweek,* April 28, 2020.

29. L. Kuo, G.J. Godeke, M.J. Raamsman, P.S. Masters, and P.J. Rottier, "Retargeting of Coronavirus by Substitution of the Spike Glycoprotein Ectodomain: Crossing the Host Cell Species Barrier," *Journal of Virology* 74, no. 3 (February 2000): 1393–1406.

30. Board on Life Sciences, Division on Earth and Life Studies, Committee on Science, Technology, and Law, Policy and Global Affairs, Board on Health Sciences Policy, National Research Council, Institute of Medicine, *Potential Risks and Benefits of Gain-of-Function Research: Summary of a Workshop* (Washington, DC: National Academies Press, 2015), 4.

31. Robert F. Kennedy Jr., *The Wuhan Cover-Up: And the Terrifying Bioweapons Arms Race* (New York: Skyhorse Publishing, 2023), Kindle Edition.

32. Kennedy, *Wuhan Cover-Up.*

33. "United States Population," *Worldometer*, accessed April 30, 2025; U.S. National Security Council, *National Security Study Memorandum NSSM 200: Implications of Worldwide Population Growth for U.S. Security and Overseas Interests* (Washington, DC: December 10, 1974), declassified July 3, 1989.

34. Ron A. M. Fouchier, "Studies on Influenza Virus Transmission between Ferrets: The Public Health Risks Revisited," *mBio* 6 (2015): e02560-14; Stephen A. Lauer et al., "The Incubation Period of Coronavirus Disease 2019 (COVID-19) From Publicly Reported Confirmed Cases: Estimation and Application," *Annals of Internal Medicine* 172, no. 9 (2020): 577–582; World Health Organization, *An R&D Blueprint for Action to Prevent Epidemics: Plan of Action* (Geneva: WHO, 2016).

35. Michaeleen Doucleff, "Why It's Difficult for Viruses to Turn into Deadly Pandemics," *NPR*, May 29, 2018.

36. John M. Barry, *The Great Influenza: The Story of the Deadliest Pandemic in History* (New York: Viking, 2004).

37. Michael Lewis, *The Premonition: A Pandemic Story* (New York: W. W. Norton & Company, 2021).

38. The White House, *National Strategy for Pandemic Influenza* (Washington, DC: November 1, 2005).

39. M. Elizabeth Halloran et al., "Modeling Targeted Layered Containment of an Influenza Pandemic in the United States," *Proceedings of the National Academy of Sciences* 105, no. 12 (March 25, 2008): 4639–4644.

40. Bernard Guyer, Mary Anne Freedman, Donna M. Strobino, and Edward J. Sondik, "Annual Summary of Vital Statistics: Trends in the Health of Americans During the 20th Century," *Pediatrics* 106, no. 6 (December 2000): 1307–1317; National Institutes of Health (NIH), "Bacterial Pneumonia Caused Most Deaths in 1918 Influenza Pandemic: Implications for Future Pandemic Planning," *NIH News Releases*, August 19, 2008.

41. National Institute of Allergy and Infectious Diseases, *Congressional Justification FY 2023*, Department of Health and Human Services, National Institutes of Health; Kennedy, *Wuhan Cover-Up*, 172.

42. "An Engineered Doomsday," *The New York Times*, January 7, 2012.

43. Rachel Nowak, "Killer Mousepox Virus Raises Bioterror Fears," *New Scientist*, January 10, 2001.

44. Robert Roos, "Think Tank Sees Big Risks in Flu Gain-of-Function Research," *CIDRAP News*, September 6, 2013.

45. Anthony S. Fauci, Gary J. Nabel, and Francis S. Collins, "A Flu Virus Risk Worth Taking," *The New York Times*, December 30, 2011.

46. Board on Life Sciences, Division on Earth and Life Studies, Committee on Science, Technology, and Law, Policy and Global Affairs, Board on Health Sciences Policy, National Research Council, Institute of Medicine, *Potential Risks and Benefits of Gain-of-Function Research: Summary of a Workshop* (Washington, DC: National Academies Press, 2015), 5.

47. Jocelyn Kaiser, "Six Vials of Smallpox Discovered in U.S. Lab," *Science*, July 8, 2014; David Malakoff, "CDC Says 75 Workers May Have Been Exposed to Anthrax," *Science*, June 19, 2014; Jocelyn Kaiser, "Lab Incidents Lead to Safety Crackdown at CDC," *Science*, July 11, 2014.

48. Jon Cohen, "Updated U.S. Biosafety Panel to Come Out of Hibernation with New Members," *Science*, July 15, 2014; Paul D. Thacker, "A Continued Candid Conversation with Richard Ebright on the History of U.S. Research Funding for Biological Agents in America and Abroad that Lack Critical Safety Overview," *Dichron Interview (Part 2)*, August 17, 2021.

49. Jocelyn Kaiser and David Malakoff, "U.S. Halts Funding for New Risky Virus Studies, Calls for Voluntary Moratorium," *Science*, October 17, 2014.

50. V. Menachery, B. Yount, K. Debbink, et al., "A SARS-Like Cluster of Circulating Bat Coronaviruses Shows Potential for Human Emergence," *Nature Medicine* 21 (2015): 1508–1513; *National Institutes of Health (NIH)*, "Gain-of-Function Deliberative Process Written Public Comments: Oct. 19, 2014 - Jun. 8, 2016," accessed April 30, 2025.

51. D. Butler, "Engineered Bat Virus Stirs Debate Over Risky Research," *Nature* (2015).

52. "NIH REPORT - Search Results for NIAID Projects," *National Institutes of Health*, accessed April 30, 2025; "NIH Lifts Funding Pause on Gain-of-Function Research," *National Institutes of Health*, December 19, 2017.

53. Sharri Markson, "How US Cash Funded Wuhan Lab Dealing in Deadly Viruses," *The Times*, September 4, 2021.

54. Keoni Everington, "WHO Inspector Caught on Camera Revealing Coronavirus Manipulation in Wuhan Before Pandemic," *Taiwan News*, January 18, 2021; Li W, Shi Z, Yu M, Ren W, Smith C, Epstein JH, Wang H, Crameri G, Hu Z, Zhang H, Zhang J, McEachern J, Field H, Daszak P, Eaton BT, Zhang S, Wang LF. "Bats Are Natural Reservoirs of SARS-Like Coronaviruses." *Science* 310, no. 5748 (October 28, 2005): 676-679.

55. J. Cui, F. Li, and Z.L. Shi, "Origin and Evolution of Pathogenic Coronaviruses," *Nature Reviews Microbiology* 17 (2019): 181–192.

56. Josh Rogin, "In 2018, Diplomats Warned of Risky Coronavirus Experiments in a Wuhan Lab. No One Listened," *The Washington Post*, March 8, 2021; "China Opens First Bio Safety Level 4 Laboratory," *U.S. Department of State*, January 19, 2018.

57. Paul D. Thacker, "The COVID-19 Lab Leak Hypothesis: Did the Media Fall Victim to a Misinformation Campaign?" *BMJ* 374 (July 8, 2021): n1656; Sharon Lerner, Mara Hvistendahl, and Maia Hibbett, "NIH Documents Provide New Evidence U.S. Funded Gain-of-Function Research in Wuhan," *The Intercept*, September 9, 2021; Jeremy Dick, "O.J. Simpson Says He's Worried He'll Run Into the 'Real' Killer If He Goes to L.A.," *MovieWeb*, August 11, 2021.

58. P. Zhou, X.L. Yang, X.G. Wang, et al., "A Pneumonia Outbreak Associated with a New Coronavirus of Probable Bat Origin," *Nature*.

59. Smriti Mallapaty, "Wuhan Lab Samples Hold No Close Relatives to Virus Behind COVID," *Nature*, December 6, 2024, corrected December 9, 2024; Nicholson Baker, "The Coronavirus Lab-Leak Theory Still Isn't Credible," *New York Magazine*; Alexandra Stevenson, "Senator Tom Cotton Repeats Fringe Theory of Coronavirus Origins," *The New York Times*, February 17, 2020.

60. "Dr. Fauci and CDC Director Walensky Testify on Efforts to Combat COVID-19," *U.S. Senate Health, Education, Labor, and*

Pensions Committee, May 11, 2021; Joseph R. Biden, *Executive Grant of Clemency: Full and Unconditional Pardon for Dr. Anthony S. Fauci,* January 19, 2025.

SUMMER

1. Christopher Nolan, *The Dark Knight,* directed by Christopher Nolan (Burbank, CA: Warner Bros., 2008). **Fair Use Disclaimer**:

The quote from *The Dark Knight* (2008) is used under the fair use provision of U.S. copyright law (17 U.S.C. § 107) for the purposes of commentary and scholarly analysis. All rights to the original work are held by the copyright holder.

2. "George Floyd Protests," *Wikipedia,* accessed April 30, 2025; Christopher C. Odom, *Justice for George Floyd: The Tipping Point?* (PhD diss., University of Central Florida, 2023).

3. Martin Luther King Jr., "I Have a Dream," *Address delivered at the Lincoln Memorial, Washington, D.C.,* August 28, 1963.

4. Harper Lee, *To Kill a Mockingbird* (New York: J.B. Lippincott & Co., 1960), 174.

5. Jon Entine, *Taboo: Why Black Athletes Dominate Sports and Why We're Afraid to Talk About It* (New York: PublicAffairs, 2000); Raisa Bruner, "How Rap Became the Sound of the Mainstream," *TIME,* January 25, 2018.

6. *Students for Fair Admissions, Inc. v. President and Fellows of Harvard College,* certiorari to the United States Court of Appeals for the First Circuit, No. 20-1199, argued October 31, 2022, decided June 29, 2023, Supreme Court of the United States.

7. Flynn, Karl. "End Affirmative Action in the Officer Corps." *Proceedings* 150, no. 7 (July 2024): 1,457; Grissom, Andrew R. "Workplace Diversity and Inclusion." *Reference & User Services Quarterly* 57, no. 4 (2018): 242–47.

8. Julia Ingram, "Clarence Thomas' Long Battle Against Affirmative Action," *The Washington Post,* May 9, 2023.

9. Gail Heriot, "The Sad Irony of Affirmative Action," *National Affairs,* no. 14 (Winter 2013).

10. Robin Zlotnick, "Professor Schools White Students on How They Would React if They Lived During Slavery," *Megaphone*, October 14, 2024.

11. The Holy Bible, *Matthew 5–7*, New Revised Standard Version.

12. Brendan Morrow, "New 'Wuthering Heights' Film Casting Sparks Backlash, Accusations of Whitewashing," *USA Today*, September 25, 2024; Jamie Burton, "Joyce Carol Oates Says Only 'Straight White Males' Are Cast as Villains Now," *Newsweek*; "43 Anti-White Commercials," *YouTube*, uploaded by Unfiltered Reality, July 28, 2017.

13. Raisa Bruner, "How Rap Became the Sound of the Mainstream," *TIME*, January 25, 2018.

14. Keith Tidman, "The Power of Language," *The Baltimore Sun*, February 29, 2024; "The Nuremberg Race Laws," *WWII National Museum*, January 7, 2025; Fenguyan Ji. *The Power of Words: Labels and Their Consequences in Mao's China (1949-1976)*. 2019; "Kim Jong Un's Orwellian Clampdown on Those Caught Using 'I Love You,' South Korean Words," *Livemint*, June 30, 2023.

15. Henry Lee, "Case Tossed Against Man Accused of San Francisco Hate Crime, Assault," *KTVU FOX 2*, March 20, 2025; *Cohen v. California*, 403 U.S. 15 (1971); Civil Rights Act of 1964, Pub. L. No. 88-352, 78 Stat. 241 (1964).

16. John Eligon, "A Debate Over Identity and Race Asks, Are African-Americans 'Black' or 'black'?" *The New York Times*, June 26, 2020.

17. Jonah Weiner, "'Niggas,' in Practice," *Slate*, June 12, 2012; Scott Neuman, "Papa John's Founder Quits as Chairman After Using the N-Word During Conference Call," *NPR*, July 12, 2018; Jonathan Capehart, "Should You Say the N-Word? No, Especially If You're Not Black," *The Washington Post*, May 7, 2021; Shannon Power, "Paris Hilton Blames 2007 N-Word Scandal on Abusive Schools She Attended," *Newsweek*, February 17, 2023.

18. "The Federal Budget in Fiscal Year 2023: An Infographic," *March 5, 2024.*

19. Patricia Hill Collins, *On Intellectual Activism* (Temple University Press, 2012), 127.

20. Percy Brown, "Why 'Color-Blindness' Is Dangerous," *Wisconsin Examiner*, December 5, 2023.

21. Neil MacMaster, *Racism in Europe 1870–2000* (St. Martin's Press, 2001); "United Nations International Criminal Tribunal for the former Yugoslavia," accessed April 30, 2025; Rotem Kowner and Walter Demel, eds., *Race and Racism in Modern East Asia* (Leiden, The Netherlands: Brill, 2015).

22. Tom Rosentiel, "Inside Obama's Sweeping Victory," *Pew Research Center*, November 5, 2008; Martin Jacques, "The Global Hierarchy of Race," *The Guardian*, September 19, 2003.

23. "Latasha Harlins | LA 92 | National Geographic," *YouTube*, posted by National Geographic Nederland-België, June 2, 2017.

24. Brenda E. Stevenson, *The Contested Murder of Latasha Harlins: Justice, Gender, and the Origins of the LA Riots* (New York: Oxford University Press, 2013).

25. Michelle Garcia, "To Many Black Americans, the O.J. Simpson Verdict Was Bigger Than O.J. Simpson," *NBC News*, April 12, 2024.

26. Manny Fernandez and Audra D. S. Burch, "George Floyd, From 'I Want to Touch the World' to 'I Can't Breathe'," *The New York Times*, April 20, 2021.

27. Associated Press, "A Long Look at the Complicated Life of George Floyd," *Chicago Tribune*, June 12, 2020.

28. Robby Soave, "Tim Walz Was a COVID-19 Tyrant," *Reason*, August 6, 2024.

29. Kevin McCoy and Grace Hauck, "George Floyd's Girlfriend Courteney Ross Gives Jurors First Glimpse of His Personal Life, Good Times and Bad," *USA Today*, April 1, 2021.

30. Virgil McDill, "U.S. Opioid Crisis Worsened During COVID-19 Pandemic," *School of Public Health, University of Minnesota*, November 30, 2023; Bridget Balch, "54 Million People in America Face Food Insecurity During the Pandemic. It Could Have Dire Consequences for Their Health," *AAMCNews*, October 15, 2020; "U.S. Experienced Highest Ever Combined Rates of Deaths Due to Alcohol, Drugs, and Suicide During the COVID-19 Pandemic," *Trust for America's Health*, May 24, 2022; Gene Falk, Isaac

A. Nicchitta, Emma C. Nyhof, and Paul D. Romero, *Unemployment Rates During the COVID-19 Pandemic* (Congressional Research Service, Library of Congress, August 20, 2021).

31. U.S. Bureau of Labor Statistics, "Unemployment Rate (UNRATE)," *FRED Economic Data*, St. Louis FED, accessed April 4, 2025; Michael H. Brenner, "Health Costs and Benefits of Economic Policy," *International Journal of Health Services*, vol. 7, no. 4, 1977, pp. 581-623; Alex Hollingsworth, Christopher Ruhm, and Kosali Simon. "Macroeconomic Conditions and Opioid Abuse." *Working Paper 23192*, National Bureau of Economic Research, February 2017.

32. Jenesse Miller, "'Like a Stick of Dynamite': USC Scholars Reflect on Legacy of 1992 L.A. Uprising and Police Beating of Rodney King," *USC News*, April 28, 2022.

33. Vladimir I. Lenin and Leon Trotsky, *The Proletarian Revolution in Russia* (1917), 160.

34. The New York Times. "How George Floyd Was Killed in Police Custody | Visual Investigations." YouTube. Published June 1, 2020.

35. Laurel Wamsley, "Passenger in George Floyd's Car Testifies in Derek Chauvin Trial," NPR, April 13, 2021; Steve Karnowski, "Scientist Testifies About Drugs Found After George Floyd Killing," Associated Press, February 10, 2022.

36. PoliceActivity. "Full Bodycam Footage of George Floyd Arrest." YouTube video, August 10, 2020.

37. Graham Kates, "George Floyd and Derek Chauvin Worked at Same Club and May Have Crossed Paths, Owner Says," *CBS News*, June 3, 2020.

38. FOX 9 Staff, "Derek Chauvin Indictment: Boy, 14, Held by Throat, Hit with Flashlight in 2017," *FOX 9 Minneapolis-St. Paul*, May 7, 2021.

39. The New York Times, "How George Floyd Was Killed in Police Custody"

40. PoliceActivity, "Full Bodycam Footage of George Floyd Arrest."

41. Andrew M. Baker, *Autopsy Report for George Floyd*, Hennepin County Medical Examiner's Office, Case No. 20-3700, May 26, 2020.

42. "Use of Force Policy in Minneapolis," *Wilson Center for Science and Justice*, Duke University School of Law, May 31, 2020.

43. Eric Levenson and Aaron Cooper, "Derek Chauvin Says He Will Not Testify at Trial and Testimony Ends. Closings Are Set for Monday," *CNN*, April 15, 2021.

44. "The Death of George Floyd: A Timeline of a Chaotic, Emotional Week in Minneapolis," FOX 9, June 1, 2020.

45. Baker, *Autopsy Report for George Floyd*

46. Amy Forliti and Steve Karnowski, "Independent Autopsy for George Floyd Contradicts Prosecutors' Findings," *PBS NewsHour*, June 1, 2020.

47. Jon Haworth, Ella Torres, and Ivan Pereira, "Floyd Died of Cardiopulmonary Arrest, Tested Positive for COVID-19, Autopsy Shows," *ABC News*, June 3, 2020.

48. Sarah Moon, "A Seemingly Healthy Woman's Sudden Death Is Now the First Known US Coronavirus-Related Fatality," *CNN*, April 24, 2020.

49. Thomas Fuller, Mike Baker, Shawn Hubler, and Sheri Fink, "A Coronavirus Death in Early February Was 'Probably the Tip of an Iceberg,'" *Berkeley Journalism* (University of California), April 22, 2020.

50. Rakesh Kochhar, "Unemployment Rose Higher in Three Months of COVID-19 Than It Did in Two Years of the Great Recession," *Pew Research Center*, June 11, 2020.

51. Katie Warren and Joey Hadden, "All the US States That Have Imposed Curfews or Called in the National Guard to Handle the George Floyd Protests," *Business Insider*, June 1, 2020.

52. Servet Günerigök, "US Governor: 'Protests No Longer About' George Floyd," *Anadolu Agency*, May 30, 2020; Graham Kates, "George Floyd and Derek Chauvin Worked at Same Club and May Have Crossed Paths, Owner Says," *CBS News*, June 3, 2020; Rilyn Eischens, "From Building Damage to Police Payouts, the Costs of Floyd's Killing Are Piling Up," *Minnesota Reformer*, November 30, 2020.

53. Donald Trump, post on X (formerly Twitter), May 28, 2020; Victoria Bekiempis, "Troops Sent to DC During George Floyd Protests Had Bayonets, Top General Says," *The Guardian*, July 3, 2020.

54. Dan Diamond, "Suddenly, Public Health Officials Say Social Justice Matters More than Social Distance," *Politico*, June 4, 2020; Joe

Concha, "CNN Ridiculed for 'Fiery But Mostly Peaceful' Caption with Video of Burning Building in Kenosha," *The Hill*, August 27, 2020.

55. KFF. "Poll: Americans Are Leaving Home More Often Now Than in April as States Ease Social Distancing Restrictions, Though Coronavirus Fears Remain." News release, June 26, 2020.

MEDIA-HYPOCHONDRIATIC COMPLEX

1. *Tomorrow Never Dies*, directed by Roger Spottiswoode (Burbank, CA: Metro-Goldwyn-Mayer, 1997).

Fair Use Disclaimer: The quote from *Tomorrow Never Dies* (1997) is used under the fair use provision of U.S. copyright law (17 U.S.C. § 107) for the purposes of commentary and scholarly analysis. All rights to the original work are held by the copyright holder.

2. Dominick Mastrangelo, "Obama: Fox News Viewers 'Perceive a Different Reality' Than Other Americans," *The Hill*, June 14, 2021.

3. Chris Nesi, "AP Takes Down 'Fact Check' Story About Ridiculous X-Rated JD Vance Hoax," *New York Post*, July 25, 2024.

4. Ross Ibbetson, "Two US Soldiers Are Killed and 12 Are Injured after Transport Vehicle Flips on Dirt Road Heading to Yukon Training Area," *Daily Mail*, October 3, 2023; Ayesha Rascoe, "Misinformation and Conspiracy Theories About Hurricane Helene Are Spreading Online," *NPR*, October 13, 2024; Shannon Bond, "The Baltimore Bridge Collapse Gave Conspiracy Theorists a Chance to Boost Themselves," *NPR*, March 27, 2024; Katie Sanders, "Fact-Checking the Wild Conspiracy Theories Related to the Attempted Trump Assassination," *PBS NewsHour*, July 15, 2024.

5. *"Alex Jones Was Right T-Shirt,"* Crowder Shop, accessed April 30, 2025; Ej Dickson, "A Video Spreading Conspiracy Theories About a Satanic Sex-Trafficking Ring Is Going Viral," *Rolling Stone*, July 15, 2021.

6. Ja'han Jones, "It's Time to Admit the Obvious: Donald Trump Sure Is Acting like a Russian Agent," *MSNBC*, February 23, 2022.

7. Clay Shirky, *Here Comes Everybody: The Power of Organizing Without Organizations* (New York: Penguin Press, 2008).

8. Snopes.com, "Fact Check."; Glenn Kessler, "Fact Checker," *The Washington Post.*

9. Alessandro Barbero, *The Battle: A New History of Waterloo* (New York: Walker & Company, 2005); Peter Hofschröer, *1815: The Waterloo Campaign, Volume 1: Wellington, His German Allies and the Battles of Ligny and Quatre Bras* (London: Greenhill Books, 1998).

10. Rani Molla, "2020 in 20 Charts," *Vox*, December 15, 2020.

11. Victoria Matlock, *A Case Study: Socialism in Venezuela* (paper presented at the HSOG 2022 Conference, "A Nation Divided? Assessing Freedom and The Rule of Law in a Post 2020 World," Liberty University Helms School of Government, February 9, 2022); Castles, Francis G. *The Social Democratic Image of Society: A Study of the Achievements and Origins of Scandinavian Social Democracy in Comparative Perspective.* London: Routledge & Kegan Paul, 1978; John E. Walsh, Adam S. Phillips, Diane H. Portis, and William

L. Chapman, "Extreme Cold Outbreaks in the United States and Europe, 1948–99," *Journal of Climate* 14, no. 12 (2001): 2642–2658; John E. Walsh, James E. Overland, Pavel Y. Groisman, et al., "Ongoing Climate Change in the Arctic," *AMBIO* 40, Suppl. 1 (2011): 6–16; Maryam Mahdavinia, Kathryn J. Foster, Esteban Jauregui, et al., "Asthma Prolongs Intubation in COVID-19," *Journal of Allergy and Clinical Immunology: In Practice* 8, no. 7 (July–August 2020): 2388–2391.

12. Eric Bradner, "Conway: Trump White House Offered 'Alternative Facts' on Crowd Size," *CNN*, January 23, 2017.

13. Wilde, Oscar. *The Decay of Lying.* Vol. 5. (New York: Doubleday, 1923).

14. *The Matrix*, directed by Lana Wachowski and Lilly Wachowski (Burbank, CA: Warner Bros., 1999), film; abragwagwa, *"Simulacra and Simulation – 'The Matrix' (1999),"* YouTube video, 3:17, April 21, 2012.

15. Jean Baudrillard, *Simulacra and Simulation*, trans. Sheila Faria Glaser (Ann Arbor: University of Michigan Press, 1994).

16. *Friends*, created by David Crane and Marta Kauffman (Burbank, CA: Warner Bros. Television, 1994–2004), television series.

17. *Saving Private Ryan*, directed by Steven Spielberg (Universal City, CA: DreamWorks Pictures, 1998), film.

18. *Suits*, created by Aaron Korsh (Universal City, CA: Universal Cable Productions, 2011–2019), television series.

19. James Poniewozik, "500 Channels and Everything's On: The Too-Much-TV Problem," *Time*, May 26, 2015; Statista, "Film Production Worldwide," accessed April 30, 2025.

20. William Shakespeare, *Macbeth*, ed. Stephen Orgel (New York: Penguin Books, 2000).

21. Jean Baudrillard, *The Gulf War Did Not Take Place*, trans. Paul Patton (Bloomington: Indiana University Press, 1995).

22. Michael R. Gordon and Bernard E. Trainor, *The Generals' War: The Inside Story of the Conflict in the Gulf* (Boston: Little, Brown and Company, 1995).

23. Adam Curtis, *HyperNormalisation*, produced by BBC (2016); Frank Dikötter, *Mao's Great Famine: The History of China's Most Devastating Catastrophe, 1959-1962* (New York: Walker & Company, 2010).

24. Carl von Clausewitz, *On War*, ed. and trans. Michael Howard and Peter Paret (Princeton, NJ: Princeton University Press, 1976), 87.

25. John Bulloch and Harvey Morris, *The Gulf War: Its Origins, History and Consequences* (1st ed.; Routledge, 1989).

26. Doug Most and Jackie Ricciardi, "Five Years Later, Scientists Study Novids—People Who've Never Had COVID," *BU Today*, March 17, 2025.

27. Rachel Lancaster, Melissa Sanchez, Kristen Maxwell, and Rachel Medley, "Original Research: TikTok's 'Dancing Nurses' During the COVID-19 Pandemic: A Content Analysis," *American Journal of Nursing* 122, no. 12 (December 1, 2022): 24–31.

28. Spee Kosloff, Gabrial Anderson, Alexandra Nottbohm, and Brandon Hoshiko, "Proximal and Distal Terror Management Defenses: A Systematic Review and Analysis," in *Handbook of Terror Management Theory*, 2019, 31–63.

29. James Cirrone, "Husband Is Found Soaked in Blood after 'Beating Wife to Death' and Dumping Her Dogs on Texas Highway," *Daily Mail*, March 17, 2025; Mithil Aggarwal, Tom Winter, Jonathan Dienst, and Myles Miller, "NYC Machete Attacker Expressed Militant Support for Islam, May Have Expected to Die, Officials Say," *NBC News*, January 1, 2023.

30. U.S. Department of Health and Human Services, *Our Epidemic of Loneliness and Isolation: The U.S. Surgeon General's Advisory on the Healing Effects of Social Connection and Community* (Washington, DC: U.S. Public Health Service, 2023).

31. Steven Pinker, *Enlightenment Now: The Case for Reason, Science, Humanism, and Progress* (New York: Viking, 2018).

32. Jacob Poushter, "How People Around the World See the U.S. and Donald Trump in 10 Charts," *Pew Research Center*, January 8, 2020.

33. Robert Warren, "US Undocumented Population Increased to 11.7 Million in July 2023: Provisional CMS Estimates Derived from CPS Data," *Center for Migration Studies*, September 5, 2024; Matthew Impelli, "Army Recruitment 'Going Like Gangbusters,' 81K Set to Join Up," *Newsweek*, January 17, 2025.

34. Edward Gibbon, *The History of the Decline and Fall of the Roman Empire*, ed. David Womersley (London: Penguin Classics, 1994).

CROSSING THE RUBICON

1. William Shakespeare, *Julius Caesar*, ed. David Daniell (London: Bloomsbury Arden Shakespeare, 1998).

2. Tom Holland, *Rubicon: The Last Years of the Roman Republic* (New York: Doubleday, 2003).

3. Nick Bryant, "The Time When America Stopped Being Great," *BBC News*, November 2, 2017.

4. "The Last Supper, Leonardo da Vinci (1452-1519)," *Cenacolo Vinciano;* "Tiznit, Cy Twombly," *Museum of Modern Art;* "The Beatles - Hey Jude," *YouTube*, video, posted by *The Beatles*, December 7, 2015; "Lil Uzi Vert - XO Tour Llif3 (Official Music Video)," *YouTube*, video, posted by *LIL UZI VERT*, September 4, 2017; "San Francisco Federal Building," *Wikipedia;* "Florence Cathedral," *Wikipedia.*

5. Arthur C. Danto, *The Transfiguration of the Commonplace: A Philosophy of Art* (Cambridge, MA: Harvard University Press, 1981).

6. Leonard Cohen, "Hallelujah," track 5 on *Various Positions*, Columbia Records, 1984. **Fair Use Disclaimer:**
The lyrics from *Hallelujah* by Leonard Cohen are quoted under the fair use provision of U.S. copyright law (17 U.S.C. § 107) for the purpose of commentary and scholarly analysis. All rights to the original work are held by the copyright holder.

7. Big Sean, "I Do It," track 3 on *Finally Famous,* produced by No I.D. and The Legendary Traxster, GOOD Music/Def Jam Recordings, 2011. **Fair Use Disclaimer:** The lyrics from *"I Do It"* by Big Sean are quoted under the fair use provision of U.S. copyright law (17 U.S.C. § 107) for the purposes of commentary and scholarly analysis. All rights to the original work are held by the copyright holder.

8. Sexyy Red and Bruno Mars, "Fat Juicy and Wet," Genius, accessed April 30, 2025; Ice Spice, "Think U The Shit (Fart)," Genius, accessed April 30, 2025.

9. Jennifer Finney Boylan, "I'm a Transgender Woman. This Is Not the Metamorphosis I Was Expecting," *The New York Times*, February 12, 2025.

10. Jonathan Lambert, "No 'Gay Gene': Massive Study Homes in on Genetic Basis of Human Sexuality," *Science*, August 29, 2019; Barry Yeoman, "Same-Sex Behavior Among Animals Isn't New. Science Is Finally Catching Up," *National Wildlife Federation*, July 4, 2023; United States Supreme Court, *Obergefell v. Hodges*, 576 U.S. 644 (2015).

11. Adam Liptak, "In Narrow Decision, Supreme Court Sides With Baker Who Turned Away Gay Couple," *The New York Times*, June 4, 2018; Clare Mulroy, "Pride Month is Here! Here's When Major Cities Celebrate with Parades in 2024," *USA Today*, April 26, 2024.

12. Jim Garamone, "Biden Administration Overturns Transgender Exclusion Policy," *DOD News*, January 25, 2021; Geoff Mulvihill, "New Federal Rule Bars Transgender School Bathroom Bans, but It Likely Isn't the Final Word," *The Associated Press*, April 23, 2024.

13. Lauren Merola, "Trump Administration Says Penn Violated Title IX by Allowing Transgender Swimmer," *The New York Times*, April 29, 2025; Grace Abels, "Harris Has Supported Gender-Affirming Care for Incarcerated People, but Trump Ads Need Context," *PolitiFact*, October 28, 2024; "The Impact of Proposed Title IX Changes on Women's Sports," *American Cornerstone Institute*, August 7, 2024.

14. Alice Motion, "'Equality Means More Than Passing Laws," *Chemistry World*, April 25, 2025.

15. C. Farmer, M. Salter, and D. Woodlock, "A Review of Academic Use of the Term 'Minor Attracted Persons,'" *Trauma Violence Abuse* 25,

no. 5 (December 2024): 4078-4089; United States Supreme Court. *Employment Division v. Smith*, 494 U.S. 872 (1990).

16. "Mental Illness," *National Institute of Mental Health.*

17. Francine Russo, "Where Transgender Is No Longer a Diagnosis," *Scientific American*, January 6, 2017.

18. François Furet, *The Peasantry of France: The Social Origins of the French Revolution* (Chicago: University of Chicago Press, 1976).

19. Editors of *Rethinking Schools*, "Celebrating Transgender Students in Our Classrooms and in Our Schools," *Rethinking Schools;* KE Pickett, OW James, and RG Wilkinson, "Income Inequality and the Prevalence of Mental Illness: A Preliminary International Analysis," *Journal of Epidemiology & Community Health* 60, no. 7 (July 2006): 646-47.

20. Sarah Clarke, "When Was the T Added to LGBT? A Quick History," *Trans Writes*, July 3, 2024.

21. Brooke Migdon, "Democrats Wrestle Over Role of Transgender Rights in Election," *The Hill*, November 18, 2024.

22. Michelle Goldberg, "The Great Capitulation," *The New York Times*, December 16, 2024.

23. Andrew Marantz, "Curtis Yarvin Thinks America Needs a King," *The New York Times*, January 18, 2025.

24. Alan Rappeport and Tony Romm, "Trump Budget to Take Ax to 'Radical' Safety Net Programs," *The New York Times*, April 25, 2025.

25. Brian Riedl, "50 Examples of Government Waste," *The Heritage Foundation*, October 6, 2009.

26. R.P. Greenberg and P.B. Zeldow, "Personality Characteristics of Men with Liberal Sex-Role Attitudes," *Journal of Psychology* 97, no. 2 (November 1977): 187-90; "PhD Dissertations and Theses | Political Economy and Government," *Harvard Kennedy School.*

27. "How Many People Receive Government Assistance?" *USAFacts;* Garrett Quinn, "Can the Libertarian Party Get 1 Percent of the Vote?" *Reason*, December 2012.

28. Matt K. Lewis, "Trump Voters Are Starting to Get Burned by DOGE's Chainsaw," *The Hill*, March 19, 2025; Tami Luhby, "Trump Has Promised to Protect Social Security. His Proposals Could Lead to Benefit Cuts in 6 Years," *CNN*, October 21, 2024.

29. Bethan Sexton, "How America Quietly Funded Iraqi Sesame Street for Years," *Daily Mail*, February 11, 2025; Rebecca Santana, "4 FEMA Employees Are Fired Over Payments to Reimburse New York City for Hotel Costs for Migrants," *The Associated Press*, February 11, 2025; Sean Burch, "DOGE Cancels Politico's Government Funding After $8M In Subscriptions Revealed," *TV NewsCheck*, February 5, 2025.

30. Stephen Losey, "Battle Over Air Force's $1,300 Coffee Cups Heats Up," *Air Force Times*, October 22, 2018; "US Congress Clears $25 Million for Democracy, Gender Programmes in Pakistan," *Business Standard*, December 23, 2020.

31. Samuel Aly, "The Gracchi and the Era of Grain Reform in Ancient Rome,"
Tenor of Our Times 6, no. 1 (2017): Article 6.

32. Alexander Phillip. *The Calendar: Its History, Structure and Improvement*. University Press, 1921.

33. Mike Sullivan, "Do You Know the True Origin Story of Daylight Saving Time?" *CBS Boston*, November 3, 2023.

34. Edward Gibbon, *The History of the Decline and Fall of the Roman Empire*, edited by David Womersley (London: Penguin Classics, 1994).

35. GT Staff Reporters, "China Among Countries with Lowest Rate of Crime, Gun-Related Cases: MPS," *Global Times*, May 27, 2024.

36. John Metcalfe, "Beijing Replaced This Huge Bridge in Only 43 Hours," *Bloomberg*, November 20, 2015; "U.S. Infrastructure's 'C' Grade from Engineers Comes with a Warning," *CBS News*, March 25, 2025.

37. Pew Research Center, "Public Trust in Government: 1958–2024," *Pew Research Center*, June 24, 2024; "Think Before You Post": The U.K. Is Now Jailing People for Social Media Comments," *Standing for Freedom Center*, August 16, 2024.

38. John P. A. Ioannidis, "Infection Fatality Rate of COVID-19 Inferred from Seroprevalence Data," *Bulletin of the World Health Organization* 99, no. 1 (January 1, 2021): 19–33F.

39. CNN, "'Screw Your Freedom': Schwarzenegger Calls Out Anti-Vaxxers," video, 00:58, August 12, 2021; Jenna Ryu, "'Screw Your Freedom': Arnold Schwarzenegger Calls Anti-Maskers 'Schmucks' in Powerful Rant," *USA Today*, August 12, 2021.

40. Ronald Pestritto, "The Birth of the Administrative State: Where It Came From and What It Means for Limited Government," *The Heritage Foundation*, November 20, 2007.

41. Patrick G. Eddington, "The PATRIOT Act Has Threatened Freedom for 20 Years," *Cato Institute*, October 21, 2021.

42. John Marini, *Unmasking the Administrative State: The Crisis of American Politics in the Twenty-First Century* (New York: Encounter Books, 2019).

43. "Ron Paul 2008 Debate On Government Agencies," *YouTube*, video, length, posted by Historyrepeats-w7b, April 5, 2025; Dudley Odell McGovney, "British Privy Council's Power to Restrain the Legislatures of Colonial America: Power to Disallow Statutes: Power to Veto," *University of Pennsylvania Law Review* 94, no. 1 (1945-1946): 59-93.

44. Mary Filardo, "Good Buildings, Better Schools: An Economic Stimulus Opportunity with Long-Term Benefits," *Economic Policy Institute*, April 29, 2008.

45. ill Rosen, "Americans Don't Know Much About State Government, Survey Finds," *Johns Hopkins University*, December 14, 2018.

46. Bruce Alpert, "George W. Bush Never Recovered Politically from Katrina," *NOLA.com | Times-Picayune*, August 28, 2015.

OUR FRIENDS ACROSS THE ATLANTIC

1. Rachel Sheffield and Robert Rector, "Air Conditioning, Cable TV, and an Xbox: What is Poverty in the United States Today?" *The Heritage Foundation*, July 19, 2011.

2. George Billias, *American Constitutionalism Heard Round the World, 1776-1989: A Global Perspective* (New York: New York University Press, 2009).

3. Guy Vanthemsche, *Belgium and the Congo, 1885–1980* (Cambridge: Cambridge University Press, 2012); Eric Williams, *Capitalism and Slavery* (Chapel Hill: University of North Carolina Press, 1944).

4. "CSS Alabama Crew of the British Isles," *American Civil War Round Table*, archived September 28, 2006; Judith Fenner Gentry, "A Confederate Success in Europe: The Erlanger Loan," *Journal of Southern History* 36, no. 2 (1970): 157–188; Don H. Doyle, *The Cause of All*

Nations: An International History of the American Civil War (New York: Basic, 2015), 8.

5. Stephen W. Sears, *Gettysburg: The Definitive History of the Battle* (New York: Houghton Mifflin Harcourt, 2003).

6. Alan Axelrod, *Miracle at Belleau Wood: The Birth of the Modern U.S. Marine Corps* (Guilford, CT: Lyons Press, 2007).

7. Reed Browning, *The War of American Independence 1775-1783* (Macmillan, 1993).

8. Mack P. Holt, *The French Wars of Religion, 1562–1629* (Cambridge: Cambridge University Press, 2005), 8.

9. William Doyle, *The Oxford History of the French Revolution*, 3rd ed. (Oxford: Oxford University Press, 2018).

10. Barrett Holmes Pitner, "Viewpoint: US Must Confront Its Original Sin to Move Forward," *BBC*, June 3, 2020.

11. Tom Shachtman, *How the French Saved America: Soldiers, Sailors, Diplomats, and the Giving of France's Greatest Gift* (New York: St. Martin's Press, 2017).

12. Jean Bethke Elshtain, "Hannah Arendt's French Revolution," *Salmagundi* no. 84 (1989): 203–13.

13. Joe Biden, "Europe Has No Moral Center!" YouTube video, posted by maxthebullfrog, December 25, 2023.

14. Greg Behrman, *The Most Noble Adventure: The Marshall Plan and the Time When America Helped Save Europe* (New York: Free Press, 2007).

15. Lucy Egginton, "Materialism and Poverty in America," *University of Manchester*, April 29, 2021.

16. Cullen S. Hendrix, "Trump's Five Percent Doctrine and NATO Defense Spending," *Peterson Institute for International Economics*, February 5, 2025.

17. "Viewpoint: Why Racism in US Is Worse Than in Europe," *BBC*, May 17, 2018; John V.C. Nye, "An Economist Looks at Europe: The Myth of Free-Trade Britain," *EconLib*, March 3, 2003.

18. Laura Clancy, "Young Adults in Europe Are Critical of the U.S. and China – but for Different Reasons," *Pew Research Center*, March 22, 2023.

19. Christina Harward, "Russia Poses Long-Term Threats to Moldova's European Integration Beyond the October Elections," *ISW Press*, October 15, 2024; Jonathan Masters and Will Merrow, "Here's How Much Aid the United States Has Sent Ukraine," *Council on Foreign Relations*, March 11, 2025.

20. Helene Cooper, Julian E. Barnes, and Eric Schmitt, "Russian Military Leaders Discussed Use of Nuclear Weapons, U.S. Officials Say," *The New York Times*, November 2, 2022.

21. Edward Kaplan, *To Kill Nations: American Strategy in the Air-Atomic Age and the Rise of Mutually Assured Destruction* (Ithaca, NY: Cornell University Press, 2015).

22. Kaplan, *To Kill Nations*.

23. Justin Bronk, "The Mysterious Case of the Missing Russian Air Force," *RUSI*, February 28, 2022; Sidharth Kaushal and Sam Cranny-Evans, "Russia's Nonstrategic Nuclear Weapons and Its Views of Limited Nuclear War," *RUSI*, June 21, 2022; Peter Paret, "Napoleon and the Revolution in War," in *Makers of Modern Strategy from Machiavelli to the Nuclear Age*, ed. Peter Paret, Gordon A. Craig, and Felix Gilbert (Princeton, NJ: Princeton University Press, 1986), 123–142.

24. Daniel Dumont, "A European View on the American Welfare State," *European Journal of Social Law*, no. 1 (March 2013).

25. Stephen Castle, "England's Health Service Is in Deep Trouble, Report Finds," *The New York Times*, September 11, 2024; Ari Shapiro, Patrick Jarenwattananon, and Manuela López Restrepo, "In Britain, It Took Just One School Shooting to Pass Major Gun Control," *NPR*, June 1, 2022.

26. Pamela Duncan and Juliette Jowit, "Is the NHS the World's Best Healthcare System?" *The Guardian*, July 2, 2018; G. Marshall, H. Newby, D. Rose, and

C. Vogler, *Social Class in Modern Britain* (1st ed.; London: Routledge, 1988).

27. Sylvie Corbet, "Wake Up and Spend More on Defense, Macron Tells Europe as Trump Takes Office," *AP News*, January 20, 2025; Patrick Marnham, *The Death of Jean Moulin: Biography of a Ghost* (New York: John Murray, 2001), 100; "France," *The Holocaust Explained*, Weiner Holocaust Library, accessed April 30, 2025.

28. Michael S. Neiberg, *When France Fell: The Vichy Crisis and the Fate of the Anglo-American Alliance* (Cambridge, MA: Harvard University Press, 2021).

29. Emily Joshu, "Do YOU Really Know What's in Your Slice? Startling Comparison of What's in US vs UK Domino's Pizza Will Shock You," *Daily Mail*, July 2, 2023; Brendan M. Lynch, "Report Reveals High Levels of Added Sugar in US Infant Formula Despite Medical Recommendations," *University of Kansas*, February 24, 2025; Gokul Sudhakaran, "Artificial Food Dyes Are Toxic: Neurobehavioral Implications in Children," *Brain and Spine* 4 (2024): 102869.

30. James Pasley, "Italy's Coronavirus Lockdown Is So Severe That You Need a Form to Prove You Have a Good Enough Reason to Be Outside," *Business Insider*, March 19, 2020; "Spain's Andalusia Rolls Out Fines of Up to €600,000 for Covid-19 'Superspreaders'," *The Local Spain*, August 5, 2020; "COVID-19: France Implements Second Lockdown to Last for Next Four Weeks," *Berry Appleman & Leiden*, November 4, 2020; Aisha Zahid, "Coronavirus: Greece Reintroduces SMS Authorisation for Movement as Country Enters Second Lockdown," *Sky News*, November 7, 2020.

31. Michael D. Shear and Donald G. McNeil Jr., "Criticized for Pandemic Response, Trump Tries Shifting Blame to the W.H.O.," *The New York Times*, April 14, 2020.

32. Craig L. Symonds, *Neptune: The Allied Invasion of Europe and the D-Day Landings* (Oxford: Oxford University Press, 2014); Oscar Guinea and Oscar du Roy, "Rules Without End: EU's Reluctance to Let Go of Regulation," *European Centre for International Political Economy (ECIPE)*, September 2024.

CONCLUSION

1. Peter Baker, "Don't Call It a Bailout: Washington Is Haunted by the 2008 Financial Crisis," *The New York Times*, March 14, 2023; W. Scott Frame, Andreas Fuster, Joseph Tracy, and James Vickery, "The Rescue of Fannie Mae and Freddie Mac," *Journal of Economic Perspectives* 29, no. 2 (2015): 25–52; Jesse Eisinger, "Why Only One Top Banker Went to Jail for the Financial Crisis," *The New York Times*, April 30, 2014.

2. Joseph R. Biden, Jr., "Executive Grant of Clemency," *U.S. Department of Justice*, January 19, 2025; Joseph R. Biden, Jr., "Executive Grant of Clemency," *U.S. Department of Justice*, December 1, 2024.

3. Nathan Schneider, "The American Empire Is in Decline—and We're Not Ready for What Comes Next," *America: The Jesuit Review*, November 5, 2019.

4. Arthur R. Kroeber, "China's Currency Policy Explained," *Brookings*, September 7, 2011; "GDP, PPP (Current International $)," *World Bank*, accessed April 30, 2025.

5. Peter Thiel with Blake Masters, *Zero to One: Notes on Startups, or How to Build the Future* (New York: Crown Business, 2014); P. Berche, "Gain-of-Function and Origin of Covid-19," *Presse Médicale* 52, no. 1 (March 2023): 104167.

6. Daniel Twining, "Abandoning the Liberal International Order for a Spheres-of-Influence World Is a Trap for America and Its Allies," *German Marshall Fund;* David E. Sanger, "Power, Money, Territory: How Trump Shook the World in 50 Days," *The New York Times*, March 11, 2025.

7. Thucydides, *History of the Peloponnesian War*, trans. Richard Crawley (London: J. M. Dent & Sons, 1910), Book 5.

8. Kelsey A. Dalrymple and Joel M. Phillips, "The Complicated Rise of Social Emotional Learning in the United States: Implications for Contemporary Policy and Practice," *Harvard Educational Review* 94, no. 3 (September 1, 2024): 337–361.

9. E.R. Montoya, D. Terburg, P.A. Bos, and J. van Honk, "Testosterone, Cortisol, and Serotonin as Key Regulators of Social Aggression: A Review and Theoretical Perspective," *Motivation and Emotion* 36, no. 1 (March 2012): 65–73; P. Gangopadhyay, M. Chawla, O. Dal Monte, and S.W.C. Chang, "Prefrontal-Amygdala Circuits in Social Decision-Making," *Nature Neuroscience* 24, no. 1 (January 2021): 5–18; William H. McNeill, *The Rise of the West: A History of the Human Community*, with a Retrospective Essay (Chicago: University of Chicago Press, 1991), 61-100; Ursula Goodenough, "Our Family Tree: Chimps, Bonobos, and Our Commonality," *NPR*, December 9, 2010.

10. Leeat Ramati-Ziber, Nurit Shnabel, and Peter Glick, "The Beauty Myth: Prescriptive Beauty Norms for Women Reflect Hierarchy-

Enhancing Motivations Leading to Discriminatory Employment Practices," *Journal of Personality and Social Psychology* (2019): 1–27; Toscano, H., Schubert, T. W., Dotsch, R., Falvello, V., & Todorov, A. (2016). Physical Strength as a Cue to

Dominance: A Data-Driven Approach. *Personality and Social Psychology Bulletin, 42*(12), 1603-1616.

11. David M. Buss, "The Evolution of Human Intrasexual Competition: Tactics of Mate Attraction," *Journal of Personality and Social Psychology* 54, no. 4 (1988): 616–628; A. Sell, A.W. Lukaszewski, and M. Townsley, "Cues of Upper Body Strength Account for Most of the Variance in Men's Bodily Attractiveness," *Proceedings of the Royal Society B: Biological Sciences* 284, no. 1869 (December 20, 2017): 20171819.

12. James Tapper, "Playground Bullies Do Prosper – and Go on to Earn More in Middle Age," *The Guardian*, March 24, 2024.

13. Dylan Matthews, "We'll Miss Globalism When It's Gone," *Vox*, March 19, 2025.

14. Peace Corps, "A School in Madagascar Gets a New Water Tank," *Peace Corps Blog.*

15. Branko Milanovic, "What Comes After Globalization?," *Jacobin*, March 28, 2023; Stephen E. Ambrose and Douglas G. Brinkley, *Rise to Globalism: American Foreign Policy Since 1938*, 9th rev. ed. (New York: Penguin Books, 2010); Peter Rutland, "Racism and Nationalism," *Nationalities Papers* 50, no. 4 (2022): 629–42.

16. H. Clarke, M.C. Stewart, and K. Ho, "Did Covid-19 Kill Trump Politically? The Pandemic and Voting in the 2020 Presidential Election," *Social Science Quarterly* 102, no. 5 (September 2021): 2194–2209.

17. Alex Thompson and Erin Doherty, "Dems' Growing Divide Over Trans Rights, DEI," *Axios*, March 10, 2025.

18. Will Durant, *The Lessons of History* (New York: Simon and Schuster, 1968).

19. Tucker Davey, "Sam Harris TED Talk: Can We Build AI Without Losing Control Over It?" *Future of Life Institute*, October 7, 2016.

20. William H. McNeill, *A History of the Western World*, 5th ed. (Chicago: University of Chicago Press, 1992), 10-49.

21. Ray Kurzweil, *The Singularity Is Near: When Humans Transcend Biology* (New York: Viking, 2005); Audrey Woods, "The Death of Moore's Law: What It Means and What Might Fill the Gap Going Forward," *MIT CSAIL Alliances*, accessed April 30, 2025.

22. Philip Bump, "No, Mr. President, Not Everyone Needs to Learn How to Code," *The Atlantic*, December 9, 2013.

23. Elon Musk, "Bootloader Tweet," *X*, March 10, 2025

24. Alexander Thomas, "Transhumanism: Billionaires Want to Use Tech to Enhance Our Abilities – the Outcomes Could Change What It Means to Be Human," *The Conversation*, January 16, 2024.

25. Victor Fiorillo, "From the Main Line to Shark Tank's Biggest Success: The Scrub Daddy Inventor," *Philadelphia Magazine*, April 11, 2025.

26. Tate Moyer, "LinkedIn Is the Worst Form of Social Media: Here's Why," *The Michigan Daily*, April 11, 2023; Horatio Alger, *Ragged Dick; or, Street Life in New York with the Boot Blacks* (Boston: Loring, 1868).

27. Richard J. Herrnstein and Charles Murray, *The Bell Curve: Intelligence and Class Structure in American Life* (New York: Free Press, 1994).

28. Donald J. Trump and Tony Schwartz, *Trump: The Art of the Deal* (New York: Ballantine Books, 1987); Nandita Bose and Jarrett Renshaw, "Billionaire Cuban Stumps for Harris as Musk Hits Trail for Trump," *Reuters*, October 18, 2024.

29. Willa Grinsfelder, "Engineers Need to Learn About the Humanities," *The Johns Hopkins News-Letter*, February 1, 2018.

30. Joseph Bernstein, "The Podcaster Asking You to Side With History's Villains," *New York Times*, April 13, 2025.

31. Jerusalem Post Staff, "Dan Bilzerian 'Would Bet Entire Net Worth That Less Than 6 Million Jews Were Killed in Holocaust'," *Jerusalem Post*, November 13, 2024.

32. Jason Lemon, "Lin Wood Claims No Planes Hit Twin Towers and Pentagon on 9/11: 'We Got Played'," *Newsweek*, October 2, 2021.

33. Louis Keene, "Joe Rogan Podcast Guest Says Israel Is Behind a Jeffrey Epstein Coverup. (And 9/11, Too.)," *Forward*, March 6, 2025.

34. Hannah Arendt, *Eichmann in Jerusalem: A Report on the Banality of Evil* (New York: Viking Press, 1963).

www.ingramcontent.com/pod-product-compliance
Lightning Source LLC
Chambersburg PA
CBHW052130270326
41930CB00012B/2822